Introduction to Optimization

Revised Second Edition

Alan Parks
Department of Mathematics
Lawrence University

Contents

Introduction

Optimization problems, traditionally called *mathematical programs* seek the maximum or minimum value of a function over a domain defined by equations and inequalities. It has become standard to call this function the *objective*.[1] Over the first part of the course we will study linear optimization problems – a class that includes a great many applications. To whet your appetite, we mention that zero-sum linear games are linear optimization problems, as are many problems of resource allocation, production mixture, project scheduling, and transportation networks. There is a definitive algorithm for solving linear programs; we will work toward a thorough understanding of this algorithm, emphasizing how particular aspects of a solution inform our understanding of the original problem beyond just having the solution in hand.

Our work on linear problems will involve a version of the *Kuhn-Tucker conditions* that are key to understanding non-linear problems. The linear version of Kuhn-Tucker will lead to the notion of duality – that a given linear problem is actually two problems in one. Furthermore, we will see how to predict what happens to the solution of a problem when small changes are made in the problem parameters.

Another main section of the course will involve non-linear optimization problems, where the emphasis is less on obtaining explicit solutions as on developing the Kuhn-Tucker conditions and the technical hypothesis under which they hold. These conditions clarify and generalize both the role of critical points and the use of the Lagrange multiplier equations in calculus. We will again address the effect of changes in problem parameters.

We will next turn to the convex problems, a class that lives in a sense between the linear and general non-linear problems. In many applied optimization problems, there are theoretical reasons for assuming convexity. When

---

[1]Notice that the *objective* of an optimization problem is a *quantity*, not a *desired outcome*. This possibly confusing usage has become standard.

the Kuhn-Tucker conditions hold in a convex problem, we necessarily have a solution; this reduces the problem to the problem of solving equations.

The theory of linear optimization is supported by linear algebra, the non-linear optimization by the multivariate calculus, and the convex programming by the theory of convex sets and functions. Attaining the needed technical underpinnings in these subjects will necessitate some serious review as well as the introduction of advanced techniques, including a version of what is called the Implicit Curve Theorem. We will address this material in three substantial chapters of the text.

As usual, you will need to study the text actively. There are problems interspersed throughout the text; some will be assigned as homework – you should do as many of them as you can in any case. Because the amount of review (especially of the linear algebra) will vary from student to student, make sure you have a thorough understanding of the reasoning given in the proofs of the main theorems.

We will use a programmed solver to do much of the computational work, always remembering how much is gained by slogging through both representative and abstract problems by hand!

CHAPTER 1

Linear Algebra

We lay out the matrix facts we need for optimization. We will begin with the matrix arithmetic covered rather briskly, and then we will veer into fun technicalities. The book $[\mathbf{3}]^1$ contains a more elementary version of this material.

1. Matrix Operations

For positive integers m, n, an $m \times n$ *matrix* is a table of numbers with m rows and n columns. We say that $m \times n$ is the *size* of the matrix. For a matrix A, its i, j-entry is denoted $A[i, j]$, for $1 \leq i \leq m$ and $1 \leq j \leq n$. We find the bracket notation $A[i, j]$ to be more readable than the commonly used subscript notation: $A_{i,j}$.

Two matrices are *equal* if they have the same size and if corresponding entries are equal (as numbers). This may seem trivial, or perhaps obvious, but it carries some important subtleties.

Because of the frequent appearance of inequalities in our work, we will find it useful to write $A \leq B$ for $m \times n$ matrices A, B when we have $A[i, j] \leq B[i, j]$ for all possible i, j. Similarly, $A \geq B$ means that each entry in A is greater than or equal to the corresponding entry in B. Notice that $A \not\leq B$ *does not imply* $A \geq B$. (Example?)

It is common in applied work to identify $\mathbb{R}^n$ with the set of $n \times 1$ matrices – or with the set of $1 \times n$ matrices. We will usually prefer to write elements of $\mathbb{R}^n$ as columns, but we will be clear in each case.

---

^{1}Boldface letters in brackets refer to entries in the bibliography on p.143.

There are three operations of matrix arithmetic. We will lay out their definitions and then give the long list of familiar properties.

Matrix addition

If A and B are $m \times n$ matrices (we are saying that A and B have the same size), then the matrix $A + B$ is defined to be $m \times n$ and

$$(A + B)[i, j] = A[i, j] + B[i, j] \quad \text{for all} \quad 1 \leq i \leq m,\ 1 \leq j \leq n$$

In other words, you add matrices entry by entry. If A and B have different sizes, then their sum is not defined.

We will write $\mathbb{O}_{m \times n}$ for the $m \times n$ matrix all of whose entries are 0. Such a matrix is called a *zero matrix*. When the size of the matrix is clear from context, we'll just write $\mathbb{O}$ for the appropriate zero matrix.

Scalar multiplication

If A is an $m \times n$ matrix and $c \in \mathbb{R}$, then the matrix $c \cdot A$ is $m \times n$ and

$$(c \cdot A)[i, j] = c \cdot \big(A[i, j]\big) \quad \text{for all} \quad 1 \leq i \leq m,\ 1 \leq j \leq n$$

Matrix multiplication

If A is $m \times n$ and B is $n \times r$, then $A \cdot B$ is defined to be $m \times r$ with

$$(A \cdot B)[i, j] = \sum_{k=1}^{n} A[i, k] \cdot B[k, j] \quad \text{for all} \quad 1 \leq i \leq m,\ 1 \leq j \leq r$$

Notice that $(AB)[i, j]$ is the *dot product* of row i of A with column j of B.

For each positive integer n, there is an $n \times n$ *identity matrix* I_n defined by the following rule. For $1 \leq i, j \leq n$, we have

$$I_n[i, j] = \begin{cases} 1 & \text{if } i = j \\ 0 & \text{if } i \neq j \end{cases}$$

In other words, I_n has 1's on the *diagonal* and 0's elsewhere.

$$I_1 = 1, \quad I_2 = \begin{pmatrix} 1 & 0 \\ 0 & 1 \end{pmatrix}, \quad I_3 = \begin{pmatrix} 1 & 0 & 0 \\ 0 & 1 & 0 \\ 0 & 0 & 1 \end{pmatrix}, \quad \dots$$

Properties of the operations

Each of the following identities follows directly from the definitions of the operations involved. We will have time for a couple of proofs in class, and you will be asked to supply a couple of proofs on homework problems. In any case, we will use these properties without reference throughout the rest of the course.

PROPOSITION 1.1. *Let A be an $m \times n$ matrix, and let B, C be matrices of the appropriate size in each case to make the stated operations defined. Let $a, b \in \mathbb{R}$. Then we have the following: (1) $A + B = B + A$;*
(2) $(A + B) + C = A + (B + C)$;   (3) $A + \mathbb{O}_{m \times n} = A$;
(4) $a \cdot (A + B) = (a \cdot A) + (a \cdot B)$;   (5) $(a + b) \cdot A = (a \cdot A) + (b \cdot A)$;
(6) $0 \cdot A = \mathbb{O}_{m \times n}$ and $1 \cdot A = A$;   (7) $I_m \cdot A = A = A \cdot I_n$;
(8) $(A \cdot B) \cdot C = A \cdot (B \cdot C)$;   (9) $a \cdot (AB) = (a \cdot A)B = A(a \cdot B)$;
(10) $A(B + C) = (AB) + (AC)$;   (11) $(A + B)C = (AC) + (BC)$

These properties are similar in many ways to properties of numbers and vectors. It is important to remember, however, that matrix multiplication is not always commutative. Here is a simple example: show that

$$\begin{pmatrix} 0 & 1 \\ 0 & 0 \end{pmatrix} \cdot \begin{pmatrix} 0 & 0 \\ 1 & 0 \end{pmatrix} \neq \begin{pmatrix} 0 & 0 \\ 1 & 0 \end{pmatrix} \cdot \begin{pmatrix} 0 & 1 \\ 0 & 0 \end{pmatrix}$$

There is no cancellation rule for multiplication. The simple example

$$\begin{pmatrix} 0 & 1 \\ 0 & 0 \end{pmatrix} \cdot \begin{pmatrix} 1 \\ 0 \end{pmatrix} = \begin{pmatrix} 0 \\ 0 \end{pmatrix}$$

shows that we can have $A \cdot B = \mathbb{O}$, and yet $A \neq \mathbb{O}$ and $B \neq \mathbb{O}$. Some matrices, however, can be canceled because they have an *inverse*. If A, B are matrices such that $A \cdot B$ and $B \cdot A$ are identity matrices, then A, B are *inverses* of each other. A matrix that has an inverse is called an *invertible matrix*. In the equation $A \cdot C = D$, if A has an inverse B, then

$$B \cdot A \cdot C = B \cdot D \quad \text{so that} \quad C = B \cdot D$$

Later, we will show that an invertible matrix has to be *square* – it has the same number of rows as columns. For now, we will prove that inverses are unique. For if B, C are both inverses of A, then using I for the identity matrix of appropriate size and using Proposition 1.1(8) (the associative law of multiplication), we see that

$$B = B \cdot I = B \cdot (A \cdot C) = (B \cdot A) \cdot C = I \cdot C = C$$

It will not surprise you that we write the unique inverse of A as A^{-1}.

◇ Problem 1

Suppose that the matrix J satisfies $J \cdot A = A$ for all $m \times n$ matrices A. Show that $J = I_m$.

◇ Problem 2

Show that if A and B invert, then so does $A \cdot B$. (Hint: use the inverses of A and B to get the inverse of $A \cdot B$, but remember that multiplication is usually not commutative.)

◇ Problem 3

Show, by finding explicit examples, that there are infinitely many matrices A such that $A^2 = I_2$.

◇ Problem 4

Suppose that A and B are $m \times n$, and that $AX = BX$ for all $n \times 1$ matrices X. Then $A = B$.

Transpose

We will need one more operation: the transpose. Given the $m \times n$ matrix A, its *transpose* A^T is the $n \times m$ matrix with $A^T[i, j] = A[j, i]$ for $1 \le i \le n$ and $1 \le j \le m$. Thus, A^T writes each row of A as a column, and each column of A as a row.

Here are the properties of the transpose; as with Proposition 1.1, the proofs will be left partly to class and partly to homework.

PROPOSITION 1.2. *Let A, B be matrices of the appropriate size in each case to make the stated operations defined. Let $a \in \mathbb{R}$. Then we have the following.*

(a) $(A^T)^T = A$
(b) $(A + B)^T = (A^T) + (B^T)$
(c) $(a \cdot A)^T = a \cdot (A^T)$
(d) $(A \cdot B)^T = (B^T) \cdot (A^T)$

◇ Problem 5

Let A be an $n \times n$ matrix, and define $f : \mathbb{R}^n \to \mathbb{R}$ by $f(X) = X^T \cdot A \cdot X$ (writing X as a column).

(a) Show that if A^T is used in place of A in the definition of f, then the same function results.

(b) Let $B = (A + A^T)/2$. Show that $B^T = B$ and that if B is used in place of A in the definition of f, then the same function results.

◇

2. Reduced Form and Replacement

Matrices will arise in this course in the context of linear equations. If A is $m \times n$ and B is $m \times 1$, the equation $A \cdot X = B$, where X is an unknown $n \times 1$ matrix, is a *linear equation* with *coefficient matrix* A and *right side* B. The reader probably knows how to use *elimination*, often called *Gaussian elimination* or *Gauss-Jordan elimination* to solve linear equations. We will describe this technique as a way of transforming a matrix into a specific form called *reduced form*. The matrix A is in *reduced form* if it has two properties:

(1) Any zero rows of A are grouped at the bottom.
(2) For each non-zero row j of A, there is a *basic column* c_j such that $A[j, c_j] = 1$ and where $A[k, c_j] = 0$ for all $k \neq j$.

Every identity matrix is in reduced form. So are all the zero matrices. Here are some other examples.

$$\begin{pmatrix} 0 & 3 & 1 \\ 1 & -4 & 0 \\ 0 & 0 & 0 \end{pmatrix}, \quad \begin{pmatrix} 1 & 1 & 1 & 0 \\ 2 & 0 & 3 & 1 \end{pmatrix}, \quad \begin{pmatrix} 3 & 4 & 1 & 1 \\ 0 & 0 & 0 & 0 \end{pmatrix}$$

The example on the far right shows that c_1 could be either 3 or 4. In other words, the columns c_j are not uniquely determined by the reduced form. When we speak of reduced form, however, we assume that particular choices of those columns have been made.[2] The *rank* of a matrix in reduced form is the number of non-zero rows it has. (Every zero matrix has rank 0.)

The main theoretical fact: for each matrix A, there is an invertible matrix R such that $R \cdot A$ is in reduced form. To prove this and to show how to find such a matrix R, we introduce the *elementary operations* that are done on the rows of a given matrix:

Elementary operations

(1) Multiplying a row by a non-zero number.
(2) Adding a multiple of one row to another (leaving the first row unchanged).
(3) Switching two rows.

The elementary operations are actually multiplication on the left by an invertible matrix.

---

[2]So, technically speaking, we are in reduced form if there is a list of distinct basic columns for which the stated properties hold.

PROPOSITION 1.3. *For each elementary operation Λ applied to matrices with m rows, there is an elementary operation Φ such that Λ and Φ undo each other. Suppose that the operation Λ is applied to the identity matrix I_m to produce the matrix E. Let A be $m \times n$. Then if Λ is applied to A, the matrix that results is $E \cdot A$.*

PROOF. To prove the first assertion, we consider the three types of operations in turn. If Λ is to multiply row i by the non-zero number α, then Φ is to multiply row i by the non-zero number $1/\alpha$. If Λ is to add α times row i to row j, then Φ is to add $-\alpha$ times row i to row j. If Λ is to switch rows i and j, then Φ is the same as Λ.

As to the matrix E, first let Λ multiply row i by $\alpha \neq 0$. Then E is the same as I_m, except that $E[i, i] = \alpha$. If A is $m \times n$, then $E \cdot A$ is the same as A in rows $j \neq i$. In row i, we see that $E \cdot A[i, j] = \alpha \cdot A[i, j]$ for each j. Thus, multiplication by E accomplishes the elementary operation.

Next suppose that we obtain E from I_m by switching rows i, j. Then $E \cdot A$ is the same as A in rows $k \notin \{i, j\}$. Row i of E is zero, except that $E[i, j] = 1$, and so

$$(E \cdot A)[i, k] = \sum_{q=1}^{m} E[i, q] \cdot A[q, k] = A[j, k]$$

so that row i of EA is row j of A. Similarly, row j of EA is row i of A, and so EA is obtained from A by switching rows i, j.

We leave it to you or to class to consider the elementary operation that adds a multiple of one row to another. ∎

The matrix E that goes with the operation Λ is said to *represent* Λ.

◇ Problem 6

Let

$$A = \begin{pmatrix} 1 & 2 & 3 \\ 4 & 5 & 6 \end{pmatrix}$$

For each of the following elementary operations, find a representing matrix E, and compute $E \cdot A$. (a) add 3 times row 2 to row 1; (b) switch rows 1 and 2.

◇

Suppose that an elementary operation Λ is applied to I_m to produce the matrix E. Applying the inverse operation Φ to E we get I_m. On the other hand, the inverse operation is accomplished by multiplying by a matrix F. Thus $F \cdot E = I_m$. Moreover, if we do Φ and then Λ, we also get I_m back again. Thus, $E \cdot F = I_m$. In other words, E, F are inverses.

$\Diamond$ Problem 7

For each of the following elementary operations, find the representing matrix E for the operation, and find the representing matrix F for the inverse operation. Show that $E \cdot F$ is an identity matrix.

(a) multiply row 2 by 6, applied to matrices with 3 rows

(b) add -2 times row 4 to row 2, applied to matrices with 4 rows.

$\Diamond$

If a sequence of operations $\Lambda_1, \dots, \Lambda_k$ is applied to an $m \times n$ matrix A, and if E_j represents Λ_j for each j, then the sequence of operations give the matrix

$$E_k \cdots E_2 \cdot E_1 \cdot A$$

Notice that $E_k \cdots E_1$ has inverse $E_1^{-1} \cdots E_k^{-1}$. Writing $R = E_1 \cdots E_k$, we have that R is invertible and $R \cdot A$ is the result of elementary operations.

The proof of our next proposition consists in applying a form of elimination to the matrix A.

PROPOSITION 1.4. *Let A be an $m \times n$ matrix. Then there is an invertible $m \times m$ matrix R such that $R \cdot A$ is in reduced form.*

PROOF. We will show that there is a sequence of elementary operations that puts A into reduced form.

If A is the zero matrix, then it is already in reduced form. If not, choose a non-zero entry $A[i, j]$. Define $c_1 = j$. Switch rows 1 and i to obtain the matrix A_1 where $A_1[1, c_1] \neq 0$. (If $i = 1$, then $A_1 = A$.) Now divide row 1 of A_1 by $A[1, c_1]$ to form the matrix A_2, and we have $A_2[1, c_1] = 1$. Next, add multiples of row 1 to the other rows to clear the entries in column c_1. Specifically, for each i with $2 \leq i \leq m$, add $-A_2[i, c_1]$ times row 1 to row i. We obtain a matrix

A_3 where $A_3[1, c_1] = 1$ and $A_3[i, c_1] = 0$ for $i \geq 2$. And A_3 was obtained from A by a sequence of elementary operations.

We continue, looking for a non-zero entry in row 2 or below. If there are non-zero entries here, we choose one $A_3[i, c_2]$. Notice that $c_2 \neq c_1$, since $A[i, c_1] = 0$. Switch rows if necessary to bring this entry to row 2, divide row 2 by the entry, and add multiples of row 2 to other rows to clear out column c_2. These operations will not affect column c_1, since that column is zero except for $A[1, c_1] = 1$.

This procedure can be applied below row 2, and then below row 3, and so on, until either we hit the bottom row or we end up with rows of 0's. The resulting matrix is in reduced form, and it was obtained from A by a sequence of elementary operations. The discussion right before this proof finds the required invertible matrix. ∎

An invertible matrix R such that $R{\cdot}A$ is in reduced form is called a *reducing matrix* for A. The matrix $R \cdot A$ is a *reduced form* for the matrix A. We will usually need merely the *existence* of a reducing matrix – we will rarely need to find it explicitly. Nonetheless, it is a good exercise to compute a couple of them to make sure you can put together the relevant facts. Proposition 1.4 shows how to perform elementary operations on an $m \times n$ matrix A to put it into reduced form, row by row. Suppose we perform the same operations in the same order starting with the identity matrix I_m, so that the matrix R results. Proposition 1.3 shows that the matrix R that results from I_m is a reducing matrix for A.

◇ **Problem 8**

In each case, find a reducing matrix for the given matrix.

a) $\begin{pmatrix} 1 & 2 & 3 \\ 4 & 5 & 6 \\ 7 & 8 & 9 \end{pmatrix}$
b) $\begin{pmatrix} 1 & 1 & 1 \\ 2 & -1 & 0 \end{pmatrix}$

Proposition 1.4 finds a particular reduced form for a given matrix. A matrix will have, in general, many reduced forms. We need to see how to change one form into another using *replacement*.

Suppose that the $m \times n$ matrix A is in reduced form, and let $A[j, c_j] = 1$ for $1 \le j \le k$, where the rows of A below row k are all 0. Suppose that $j \le k$ and $A[j, p] \ne 0$ for some $p \ne c_j$. We can perform elementary row operations to get a new reduced form in which $A[j, p]$ replaces $A[j, c_j]$. Here's what we do:

(1) Divide row j by $A[j, p]$ to form the matrix A_1 with $A_1[j, p] = 1$.
(2) Add multiples of row j to the other rows to form the matrix A_2 with $A_2[i, p] = 0$ for all $i \ne j$. (Specifically, add $-A_1[i, p]$ times row j to row i, for each $i \ne j$.)

Notice that in the matrix A_2 that results, we have $A_2[q, c_q] = 1$ for all $q \ne j$, and column c_q of A_2 is still 0 except for the entry at row q. This is because the row operations just described do not affect column c_q. We describe the process of going from A to A_2 as *replacement* and we say we *replace column c_j by column p*.

◇ **Problem 9**

Describe and make all possible replacements in the following reduced form. (Start over with the matrix given here each time.)

$$\begin{pmatrix} 1 & 0 & 2 & 1 \\ -1 & 1 & 3 & 0 \\ 0 & 0 & 0 & 0 \end{pmatrix}$$

3. Linear Equations

Now suppose we have a linear equation $A \cdot X = B$, where A is $m \times n$ and B is $m \times 1$. Proposition 1.4 finds a reducing matrix R for A. We claim that the equation $RA \cdot X = RB$ has exactly the same solutions as the equation

$AX = B$. Indeed, if $AX = B$, then certainly $RAX = RB$. Conversely, if $RAX = RB$, then since R is invertible, we have

$$AX = R^{-1}RAX = R^{-1}RB = B$$

As an alternative to finding a reducing matrix, we can simply perform elementary operations on both A, B to bring the A-part into reduced form. As we mentioned above, such a process constitutes *elimination* in its various forms.

Now we are ready to see the utility of reduced form: the solutions (or lack of solutions) of $A \cdot X = B$ are obvious if A is in reduced form. To see this, choose columns $c_1, c_2, \ldots, c_r$, where r is the rank of A, such that $A[j, c_j] = 1$ and $A[j, c_j]$ is the only non-zero entry in column c_j. Suppose first that B has a non-zero entry in a row $i > r$. Row i of A is all 0's, and so the i-th entry of $A \cdot X$ is 0. There is no way this can be equal to $B[i, 1] \neq 0$, and so the equation $A \cdot X = B$ has *no solutions*. We say that the equation is *inconsistent*.

Next suppose that $B[i, 1] = 0$ for all $i > r$. The i-th entries of $A \cdot X$ and B are 0 no matter what X is, and so we can drop these irrelevant equations. Let F be the set of columns k, other than the columns $c_1, \ldots, c_r$. The variables $X[i]$ for $i \in F$ are called the *free variables*. We claim that for each arbitrary choice of free variables $X[i]$, there is a unique solution to $A \cdot X = B$. Indeed, choose $j \leq r$ and consider the j-th entry of $A \cdot X$. We have $A[j, c_j] = 1$, so $X[c_j]$ occurs here. No other $X[c_i]$ occurs there, since $A[j, c_i] - 0$ when $i \neq j$. Thus, the j-th entry of $A \cdot X$ has this form.

$$X[c_j] + \sum_{i \in F} A[j, i] \cdot X[i]$$

The j-th entry of the equation $A \cdot X = B$ is then

(1.1) $$X[c_j] + \sum_{i \in F} A[j, i] \cdot X[i] = B[j, 1]$$

The equation shows that the choice of $X[i]$ for $i \in F$ determines this variable. Furthermore, since $X[c_j]$ occurs in no other equation, the value of $X[c_j]$ determined here will not clash with the information in any other equation.

The variables $X[c_j]$, as $1 \leq i \leq r$, are called *basic variables*.

◇ **Problem 10**

In the following reduced form, identify basic variables and free variables, and write the equations (1.1) explicitly.

$$\begin{pmatrix} 0 & 3 & 1 & 0 & 4 \\ 1 & -1 & 0 & 0 & 1 \\ 0 & 2 & 0 & 1 & -5 \\ 0 & 0 & 0 & 0 & 0 \end{pmatrix} \cdot X = \begin{pmatrix} 2 \\ 4 \\ 1 \\ 0 \end{pmatrix}$$

There is a particular solution associated with the equations (1.1). If the free variables are set to 0, and then if $X[c_j] = B[j, 1]$ for each j, we get a solution called the *basic vector* for $A \cdot X = B$. Observe that the phrase *basic vector* implies that the coefficient matrix A is in reduced form.

We put all this together.

◇ **Problem 11**

For each of the following linear equations, do the following.

    11.1 Perform elementary operations on the coefficient matrix and right side simultaneously to put the coefficient matrix in reduced form.

    11.2 From reduced form, deduce whether there are or are not solutions to the equation.

    11.3 If there are solutions, find the basic vector associated with the reduced form.

a) $\begin{pmatrix} 1 & 2 & 3 \\ 4 & 5 & 6 \\ 7 & 8 & 9 \end{pmatrix} \cdot X = \begin{pmatrix} 2 \\ 5 \\ 8 \end{pmatrix}$ b) $\begin{pmatrix} 1 & 1 & 2 & -3 \\ 2 & 1 & 3 & 2 \\ 1 & 1 & 1 & 1 \\ 0 & 1 & -2 & 4 \end{pmatrix} \cdot X = \begin{pmatrix} 2 \\ 7 \\ 7 \\ 9 \end{pmatrix}$

c) $\begin{pmatrix} 1 & 1 & 2 & -3 \\ 2 & 1 & 3 & 2 \\ 1 & 1 & 1 & 1 \\ 0 & 1 & -2 & 4 \end{pmatrix} \cdot X = \begin{pmatrix} 2 \\ 7 \\ 7 \\ 12 \end{pmatrix}$ d) $\begin{pmatrix} 1 & 2 \\ 3 & 4 \end{pmatrix} \cdot X = \begin{pmatrix} 5 \\ 6 \end{pmatrix}$

Here is a summary of what we have done in this section. If we are given an arbitrary linear equation $A \cdot X = B$, we can find a reducing matrix R for A, and then the equation $R \cdot A \cdot X = R \cdot B$ has a reduced coefficient matrix and the same solutions as the original equation. The reduced form allows us to observe whether the equation has any solutions at all. If it does, then we can write the basic variables in terms of the free variables to obtain the set of solutions. It should be noted that if there are no free variables at all, then the equation has a unique solution. We have also described *replacement* that gets us from one reduced form to another; replacement will play a key role in the solution of linear optimization problems, as we will see in Chapter 2.

4. Inverses

As an application of what we have just done, and because we are interested in the topic for its own sake, we prove the following.

PROPOSITION 1.5. *Let A be an $m \times n$ matrix. If A has an inverse, then A is square, so that $m = n$.*

PROOF. We first assume that A is in reduced form; let r be the rank. We have $r \leq m$; we claim that $r = m$. If $r < m$, choose an $m \times 1$ matrix B such that $B[m, 1] = 1$. Then the equation $A \cdot X = B$ is inconsistent. This is contradicted by the fact that $A^{-1} \cdot B$ is a solution:

$$A \cdot (A^{-1} \cdot B) = (A \cdot A^{-1}) \cdot B = B$$

Thus, $r = m$.

Next we claim that $r = n$. To see this, consider the equation $A \cdot X = \mathbb{O}_{m \times 1}$. This equation is consistent, since $X = \mathbb{O}_{n \times 1}$ is a solution. And this is the only solution, since if $AX = \mathbb{O}$, then $A^{-1} \cdot AX = A^{-1} \cdot \mathbb{O}$, so that $X = \mathbb{O}$. Since the solution is unique, there are no free variables in reduced form. Thus, the rank of reduced form must be n, the number of variables. We have proved that $m = r = n$.

We were assuming that A is in reduced form. In general, Proposition 1.4 finds an invertible matrix R such that $R \cdot A$ is in reduced form. The matrix

$R \cdot A$ is invertible, having inverse $A^{-1} \cdot R^{-1}$ (that was a problem given above!). By the first part of the proof, the matrix $R \cdot A$ is square. The matrix A has the same size, so it's square. ∎

Having an inverse is equivalent to several conditions.

PROPOSITION 1.6. *Let A be an $n \times n$ matrix. Then the following are equivalent.*

(a) A has an inverse;
(b) for each $n \times 1$ matrix B, the equation $A \cdot X = B$ has a unique solution;
(c) the only solution to $A \cdot X = \mathbb{O}_{n \times 1}$ is $\mathbb{O}_{n \times 1}$;
(d) a reduced form of A has rank n;
(e) for each $n \times 1$ matrix B, the equation $A \cdot X = B$ is consistent;
(f) there is an $n \times n$ matrix C such that $A \cdot C = I_n$.

PROOF. (a)⇒(b): The matrix $A^{-1} \cdot B$ is a solution to $A \cdot X = B$. If C is a solution to this equation, then $A^{-1} \cdot A \cdot C = A^{-1} \cdot B$, and this shows that $C = A^{-1} \cdot B$, so that the solution is unique.

(b)⇒(c): The matrix $\mathbb{O}_{n \times 1}$ is a solution to $A \cdot X = \mathbb{O}_{n \times 1}$. By (b), this is the unique solution.

(c)⇒(d): The equation $A \cdot X = \mathbb{O}$ is consistent, since it has $\mathbb{O}$ as a solution. The solutions to this equation are available from a reduced form A' for A. The equations (1.1) expressing the basic variables in terms of the free variables show that if there are free variables, there will be infinitely many solutions. That there is a unique solution shows that the rank is n.

(d)⇒(e): If R is a reducing matrix for A such that $R \cdot A$ has rank n, then we know that $A \cdot X = B$ has the same solutions as $R \cdot A \cdot X = R \cdot B$. Since the rank of $R \cdot A$ is n, there are no rows of 0's in the reduced form, and therefore, the equation is consistent.

(e)⇒(f): Let J_i be the i-th column of I_n, for $1 \le i \le n$. The equation $A \cdot X = J_i$ is consistent; let C_i be a solution. Let C be the $n \times n$ matrix having C_i as its i-th column, and you should compute that $A \cdot C = I_n$.

(f)$\Rightarrow$(a): We suppose that $A \cdot C = I_n$. Then the equation $C \cdot X = \mathbb{O}$ has a unique solution, for if $C \cdot V = \mathbb{O}$, where V is $n \times 1$, then $A \cdot C \cdot V = A \cdot \mathbb{O}$, so that $V = \mathbb{O}$. Thus, C satisfies (c); by the part of the proof already done, we see that C satisfies (f): there is an $n \times n$ matrix D such that $C \cdot D = I_n$. Compute

$$A = A \cdot I_n = A \cdot (C \cdot D) = (A \cdot C) \cdot D = I_n \cdot D = D$$

Since $A = D$, we have $A \cdot C = I_n = C \cdot A$, and A has an inverse. ■

◇ Problem 12
Show that the matrix $\begin{pmatrix} a & b \\ c & d \end{pmatrix}$ has an inverse if and only if $a \cdot d - b \cdot c \neq 0$.

◇

◇ Problem 13
Define

$$A = \begin{pmatrix} 1 & 2 & 3 \\ 4 & 5 & 6 \end{pmatrix}$$

Prove each of the following two statements by solving a system of linear equations.

(a) There is a 3×2 matrix C such that $A \cdot C = I_2$.
(b) There is no 3×2 matrix D such that $D \cdot A = I_3$.

◇

◇ Problem 14
Let A be an $m \times n$ matrix. Show that there is an $n \times m$ matrix C such that $A \cdot C = I_m$ if and only if a reduced form of A has rank m. (Hint: if C exists, then $A \cdot X = B$ is always consistent; if a reduced from has rank m, solve $A \cdot C = I_m$ for one column at a time.)

◇

5. Technical Lemmas

We collect miscellaneous facts needed later.

First, the *rank* of every reduced form of a given matrix is the same. This is sometimes called the *Rank Theorem*.

RANK THEOREM. *Let A be an $m \times n$ matrix. Then every reduced form for A has the same rank.*

PROOF. Let A_1 and A_2 be reduced forms for A, and we imagine using each of them to solve the equation $A \cdot X = \mathbb{O}$. We will use a form like (1.1) for each of the reduced forms.

Let X_1 be the column of basic variables common to A_1 and A_2; let X_2 be the basic variables in A_1 that are free in A_2; let X_3 be the free variables in A_1 that are basic in A_2; let X_4 be the free variables in both A_1 and A_2. (Some of these sets of variables may be empty; we can get by with considering only two cases, as you will see.) We suppose that the two reduced forms are these, as in (1.1), with the form of A_1 on the left and that of A_2 on the right:

$$X_1 + B_3 \cdot X_3 + B_4 \cdot X_4 = \mathbb{O} \qquad X_1 + D_2 \cdot X_2 + D_4 \cdot X_4 = \mathbb{O}$$
$$X_2 + C_3 \cdot X_3 + C_4 \cdot X_4 = \mathbb{O} \qquad X_3 + E_2 \cdot X_2 + E_4 \cdot X_4 = \mathbb{O}$$

Case 1. There are no variables in X_3 or there are no variables in X_2.

Assume there are no variables in X_3. (The case of X_2 is similar.)

Then A_1's second equation says that X_4 determines X_2, but in A_2, the variables X_2, X_4 are independent. This is a contradiction unless X_2 has no variables. Now X_1 holds all the basic variables for both A_1 and A_2, and so their rank is the same.

Case 2. There are variables in X_3 and X_2.

We substitute X_3 from the second equation from A_2 into the second for A_1 to obtain

$$(I - C_3 \cdot E_2) \cdot X_2 + (C_4 - C_3 \cdot E_4) \cdot X_4 = \mathbb{O}$$

In A_2, the variables X_2, X_4 are independent. It follows that $I = C_3 \cdot E_2$ (and $C_4 = C_3 \cdot E_4$, but we don't need this).

Similarly, substituting X_2 from the second equation in A_1 into the second for A_2, we obtain that $I = E_2 \cdot C_3$. We conclude that E_2, C_3 are inverses, and so by Proposition 1.5 they are both $k \times k$ for some k. This says that X_2 and X_3 have the same number of variables, and now we see that A_1 and A_2 have the same number of basic variables – the same rank. ■

We will need the following fact when we encounter Lagrange multipliers.

LEMMA 1.7. *Let A be $m \times n$, let B be $1 \times n$, and suppose that whenever V is $n \times 1$ with $A \cdot V = \mathbb{O}$, then we have $B \cdot V = \mathbb{O}$. Then there is an $1 \times m$ matrix C such that $B = C \cdot A$.*

PROOF. We have stated the Lemma in the way in will be used. We will give the proof in a form that is easier to understand; at the end of the argument we will derive the desired conclusion.

Let E be an $n \times m$ matrix and let F be $n \times 1$ and suppose that whenever W is $1 \times n$ and $W \cdot E = \mathbb{O}$, then $W \cdot F = 0$ as well. We will prove that the equation $E \cdot X = F$ is consistent.

If $E = \mathbb{O}_{n \times m}$, then $W \cdot E = \mathbb{O}$ for all $1 \times n$ matrices W. Then $W \cdot F = 0$ for all W. It follows that $F = \mathbb{O}$, and so $E \cdot \mathbb{O} = F$. Thus, $E \cdot X = F$ is consistent.

Now let $E \neq \mathbb{O}$. Multiply on the left by a reducing matrix G for E, and suppose that the rank is r. If $r = n$ (so that every row of E got a basic variable), then $E \cdot X = F$ is consistent. We can assume that $r < n$, so that the last $n - r$ rows of $G \cdot E$ are 0. If $E \cdot X = F$ is *not* consistent, then $(G \cdot F)[j] \neq 0$ for some $j > r$. Let W be $1 \times n$ with $W[j] = 1$ and all other entries 0. Then $W \cdot G \cdot E = \mathbb{O}$, since the first r entries of W are 0, and the last $n - r$ rows of $G \cdot E$ is 0. By hypothesis, $W \cdot G \cdot F = \mathbb{O}$, but $W \cdot (G \cdot F) = (G \cdot F)[j] \neq 0$. This contradiction proves that $E \cdot X = F$ is consistent.

Apply this reasoning with $E = A^T$ and $F = B^T$, using the matrices A, B of the hypothesis of this proposition. There is a matrix C such that $A^T \cdot C = B^T$, and we see that $C^T \cdot A = B$, as needed. ■

You are familiar with the *dot product* of vectors. If $v, w \in \mathbb{R}^n$, then

$$v \circ w = \sum_{j=1}^{n} v[j] \cdot w[j]$$

If we regard v, w as $n \times 1$ matrices, then notice that $v \circ w = v^T \cdot w$, where the right hand product is matrix multiplication. You should know that the dot product is symmetric, linear in both vectors, and that it commutes with scalar multiplication.

We will also need the concept of *norm* of matrices in general and vectors in particular. Given a vector $v \in \mathbb{R}^n$, its *norm* is

$$|v| = \sqrt{\sum_{k=1}^{n} v[k]^2}$$

If we write v as an $n \times 1$ matrix, we see that $|v|^2 = v \circ v = v^T \cdot v$.

An $m \times n$ matrix M has $m \cdot n$ entries, and so it can be considered as an element of $\mathbb{R}^{m \cdot n}$. Then we have a natural *norm*:

$$|M| = \sqrt{\sum_{i=1}^{m} \sum_{j=1}^{n} M[i,j]^2}$$

When M is $m \times 1$ or $1 \times n$, its matrix norm is obviously the same as its vector norm.

We will need the following fact – the *Cauchy-Schwarz Inequality*.

PROPOSITION 1.8. *Let* $x, y \in \mathbb{R}^n$. *Then*

$$|x \circ y| \leq |x| \cdot |y|$$

PROOF. If $|y| = 0$, then $y = \mathbb{O}$ and both inequalities are trivial. We assume that $|y| \neq 0$.

Let t be an arbitrary real number, and we consider the expression $|x - t \cdot y|^2$.

$$|x - t \cdot y|^2 = \sum_{j=1}^{n} (x[j] - t \cdot y[j])^2 = \sum_{j=1}^{n} \left[x[j]^2 - 2 \cdot t \cdot x[j] \cdot y[j] + t^2 \cdot y[j]^2 \right]$$

$$= \sum_{j=1}^{n} x[j]^2 - 2 \cdot t \cdot \sum_{j=1}^{n} x[j] \cdot y[j] + t^2 \cdot \sum_{j=1}^{n} y[j]^2$$

$$= |x|^2 - 2 \cdot t \cdot x \circ y + t^2 \cdot |y|^2$$

The last expression obtained is a quadratic function of t, and since $|y| > 0$ its graph is concave up. Its minimum occurs where the derivative (in t) is zero; that is to say at $t = x \circ y / |y|^2$. Since the expression we started with is a square, it is non-negative, and so

$$0 \leq |x|^2 - 2 \cdot \frac{x \circ y}{|y|^2} \cdot x \circ y + \left(\frac{x \circ y}{|y|^2} \right)^2 \cdot |y|^2$$

Clearing $|y|^2$ we get

$$0 \leq |x|^2 \cdot |y|^2 - (x \circ y)^2 \quad \text{which is} \quad (x \circ y)^2 \leq |x|^2 \cdot |y|^2$$

Taking the square root of both sides we obtain the Cauchy-Schwarz inequality.

$$(1.2) \qquad\qquad |x \circ y| \leq |x| \cdot |y|$$

Here is an application to the matrix norm.

LEMMA 1.9. *Let A be an $m \times n$ matrix and let B be $n \times k$. Then*

$$|A \cdot B| \leq |A| \cdot |B|$$

PROOF. Let A_i denote the i-th row of A and B_j the j-th column of B. Then the Cauchy-Schwarz inequality shows that

$$(A \cdot B)[i, j] = A_i \circ B_j \leq |A_i| \cdot |B_j|$$

and so

$$|A \cdot B|^2 = \sum_{i,j} \left((A \cdot B)[i,j] \right)^2 \leq \sum_{i,j} |A_i|^2 \cdot |B_j|^2$$

$$= \sum_{i=1}^{m} |A_i|^2 \cdot \sum_{j=1}^{k} |B_j|^2$$

The definition of the norm of A shows that

$$\sum_{i=1}^{m} |A_i|^2 = |A|^2$$

Similarly, the sum of $|B_j|^2$ is $|B|^2$, and we have $|A \cdot B|^2 \leq |A|^2 \cdot |B|^2$, and this implies the result. ∎

Define the *sphere* in $\mathbb{R}^n$ to be

$$S_n = \left\{ v \in \mathbb{R}^n \mid |v| = 1 \right\}$$

If $w \in \mathbb{R}^n$, then $w = |w| \cdot v$ for some $v \in S_n$.

Let A be an $m \times n$ matrix. For $v \in S_n$, we have $0 \leq |A \cdot v| \leq |A|$, using Lemma 1.9 for the second inequality. It follows that we can define

$$\mu(A) = \inf \left\{ |A \cdot v| \mid v \in S_n \right\} \quad \text{and} \quad \nu(A) = \sup \left\{ |A \cdot v| \mid v \in S_n \right\}$$

(Later we will see that μ and ν are actually the minimum and maximum, respectively, of their sets.)

Lemma 1.9 shows that

$$\mu(A) \cdot |v| \leq |A \cdot v| \leq \nu(A) \cdot |v| \quad \text{for all} \quad v \in \mathbb{R}^n$$

Indeed, this is true for $v = \mathbb{O}$. If $v \neq \mathbb{O}$, then $v = |v| \cdot w$, for some $w \in S_n$, and then

$$|A \cdot v| = |v| \cdot |A \cdot w|$$

and we have both inequalities.

PROPOSITION 1.10. *Let A be an $m \times n$ matrix and let B be $n \times k$. Then*

$$\mu(A) \cdot |B| \leq |A \cdot B| \leq \nu(A) \cdot |B|$$

PROOF. Let B_j denote the j-th column of B, and then

$$|A \cdot B|^2 = |\begin{bmatrix} AB_1 & AB_2 & \cdots & AB_k \end{bmatrix}|^2 = \sum_{j=1}^{k} |AB_j|^2$$

$$\leq \sum_{j=1}^{k} \nu(A)^2 \cdot |B_j|^2 = \nu(A)^2 \cdot |B|^2$$

Similarly, we obtain the lower bound. ∎

It is noteworthy that μ and ν work by pulling the matrix off from the left. If we want to pull off from the right, we can use the transpose. Indeed, given A, B such that $A \cdot B$ is defined, we can compute

$$|B^T \cdot A^T| \leq \nu(B^T) \cdot |A^T| = \nu(B^T) \cdot |A|$$

On the other hand, the norm of a matrix is equal to the norm of its transpose, and so

$$|A \cdot B| = |(A \cdot B)^T| = |B^T \cdot A^T| = |A| \cdot \nu(B^T)$$

The formula $|A \cdot B| = |A| \cdot \nu(B^T)$. shows how to pull B off from the right. A similar calculation works for μ.

PROPOSITION 1.11. *Let A be an $n \times n$ matrix. Then A is invertible if and only if $\mu(A) > 0$.*

PROOF. If A is invertible, then for all $v \in S_n$, we have

$$1 = |v| = |A^{-1} \cdot A \cdot v| \leq \nu(A^{-1}) \cdot |A \cdot v|$$

It follows that $|A \cdot v| \geq 1/\nu(A^{-1})$, and so $\mu(A) > 0$.

If A is not invertible, then there is $v \in \mathbb{R}^n$ with $v \neq \mathbb{O}$ and $A \cdot v = \mathbb{O}$. We can choose such a v in S_n, and we see that $|A \cdot v| = 0$, so that $\mu(A) = 0$. ∎

Matrices close to an invertible matrix are, themselves, invertible.

PROPOSITION 1.12. *Let A be an invertible $n \times n$ matrix. $0 < \epsilon < \mu(A)$. Let B be $n \times n$ with $|A - B| < \epsilon$. Then B is invertible and*

$$\mu(A) - \epsilon \leq \nu(B) \leq \nu(A) + \epsilon \quad \text{and} \quad \nu(B^{-1}) \leq \frac{1}{\nu(A) - \epsilon}$$

PROOF. Let B be $n \times n$ with $|A - B| < \epsilon$, and let $v \in S_n$.

$$|B \cdot v| = |A \cdot v + (B - A) \cdot v| \geq |A \cdot v| - |(B - A) \cdot v|$$
$$\geq |A \cdot v| - |B - A| \cdot |v| > \mu(A) - \epsilon$$

We see that $\mu(B) \geq \mu(A) - \epsilon > 0$. By Proposition 1.11, the matrix B is invertible.

Also,

$$|B \cdot v| \leq |A \cdot v| + |(B - A) \cdot v| \leq \nu(A) + \epsilon$$

so that $\nu(B) \leq \nu(A) + \epsilon$.

For $v \in S_n$, compute

$$1 = |v| = |B \cdot B^{-1} \cdot v| \geq (\mu(A) - \epsilon) \cdot |B^{-1} \cdot v|$$

This proves the final inequality. ∎

◇ Problem 15

Let A be an invertible $n \times n$ matrix. Let $0 < c < \mu(A)$. Show that there is $\epsilon > 0$ such that if B is $n \times n$ and $|B - A| < \epsilon$, then $\mu(B) \geq c$.

◇

◇ Problem 16

Let A be $n \times n$. Show that $\nu(A^k) \leq \nu(A)^k$ for $k = 1, 2, 3, \ldots$.

◇

◇ Problem 17

Let A be $n \times n$ with $\nu(A) < 1$. Show that this series converges:

$$\sum_{k=0}^{\infty} A^k$$

(As you might expect, we define $A^0 = I_n$.) (Hint: let E be a column of I_n, and show that the series for $A^k \cdot E$ converges, using the definition of ν.)

◇

CHAPTER 2

Linear Optimization

There are many, many application problems that are *linear programs*. One of the best references is Dantzig's very well-written classic [**1**]. Dantzig discovered the *simplex algorithm* which resolves all linear programs; the utility of this algorithm is one of the great success stories of applied mathematics in the late 1900's.

1. Definitions and Basic Solutions

We want to define what we mean by a *linear program*. Linear programs are very common, and they occur in a variety of forms. Thus, there are many possible ways to give a precise definition. We will *start* with a very general definition, making it easy to see that there are many examples; then we will move toward the two standard forms that will figure into our theoretical work.

Here is the general definition: A *linear program* (an *LP*) is an optimization problem in which the *objective*[1] is a linear function of the other variables,[2] and such that the constraints on these other variables constitute a system of linear equations and non-strict linear inequalities. We will discuss the following examples in class. This is only the beginning!

(a) Minimize $a+b-c+7$ such that $a \leq 0$, $b \geq 0$, $c \leq 10$, and $a+3b+c = 10$.
(b) Are there a, b, c such that the following conditions hold? $a+b+c \leq 10$, $2a + 3b + c = 12$, $a \geq 2$, $b, c \geq 0$.
(c) Define $f(a, b, c) = \min\{a + 2b + c, -a + 3b + 5c\}$. Find the maximum of $f(a, b, c)$ such that $a + 5b + 6c \leq 10$ and $a, b, c \geq 0$.

---

[1]Recall that the objective is the quantity to be optimized.
[2]These other variables are often called *problem variables*.

If the objective is Z and if the vector V holds values of the problem variables, we will write $Z(V)$ for the value of Z when the problem variables are set to the values indicated in V. Continuing this idea, if V satisfies the other constraints of the problem (setting the objective aside), then it is said to be a *feasible vector*.

We consider some very general possibilities for a linear program. It may be that there are no feasible vectors. Here is a trivial example.

$$\text{minimize} \quad a \quad \text{such that} \quad \begin{array}{c} a + b \leq 2 \\ a + b \geq 4 \end{array} \quad \text{with} \quad a \geq 0,\ b \geq 0$$

When there are no feasible vectors, we say that the linear program is *infeasible* and we say that it *has no solution*.

A *feasible* linear program has a non-empty set of feasible vectors. It may be that the objective Z does not have an optimum on this set. Trivial example.

$$\text{minimize} \quad x \quad \text{with} \quad x \leq 1$$

If Z does not have the desired optimum, we say that the linear program *has no solution*.

It remains that the program is feasible and Z has an optimum value. This value is unique. However, there may be several feasible vectors at which this extreme occurs. Each of these vectors is called a *solution* to the linear program. To repeat: a solution to a linear program is a feasible vector at which the objective is optimized. Here is an example of a linear program having more than one solution.

$$\text{minimize} \quad x \quad \text{such that} \quad x + y \geq 7 \quad \text{and} \quad x \geq 0,\ y \geq 0$$

◇ **Problem 18**

One of the following linear programs is infeasible, one is feasible but has no solution, one has infinitely many solutions, one has a unique solution. Figure out which is which and give convincing reasons. (Hint: it might help to graph the constraints in the x, y-plane!)

(a) Minimize $2 \cdot x + 4 \cdot y$ such that $x + 2 \cdot y \geq 3$ and $4 \cdot x + 5 \cdot y \leq 20$ and $x, y \geq 0$.

(b) Minimize $-x + 2 \cdot y + 2$ such that $-3 \cdot x + y \leq 5$ and $x, y \geq 0$.
(c) Maximize $x + 2 \cdot y$ such that $2 \cdot x + 3 \cdot y \leq 6$ and $x, y \geq 0$.
(d) Maximize $3 \cdot x + 5 \cdot y$ such that $-x + y \geq 2$ and $2 \cdot x + y \leq 1$ and $x, y \geq 0$.

$\diamondsuit$

Now we introduce our first standard form for a linear program: *primal form*. Suppose we have n variables in an $n \times 1$ matrix X. Then primal form looks like this:

$$(2.1) \qquad \text{minimize} \quad Z \quad \text{where} \quad Z + C \cdot X = z_0, \quad A \cdot X \leq B, \quad X \geq \mathbb{O}$$

The matrices A, B, C and the number z_0 are constant.[3] Notice that the $1 \times n$ *objective coefficient matrix* C is written on the same side of the equation as the objective Z. This expresses Z as one variable among many in a system of linear equations. This form will play a significant role theoretically. We will use Excel[4] to do our automated calculations; in Excel the objective is usually computed in function form: $Z = C \cdot X + z_0$ (this C is the negative of the one that occurs in (2.1)!). The matrix A is the *coefficient matrix*; if it is $m \times n$, then $AX \leq B$ stands for m inequalities in the n problem variables in X. It follows that B is $m \times 1$, it is called the *right side matrix*.

We will need to be able to put an arbitrary linear program into primal form. First, if in the given linear program, the objective Z is to be *maximized*, then we can change the objective to $W = -Z$. If the minimum of W occurs when $W = 4$, say, then the maximum of Z is -4. Second, notice that the problem variables need to be non-negative. An unconstrained variable x can be written $x = a - b$, where a, b are non-negative variable. We can substitute $a - b$ for x wherever it occurs. A problem solution gives us values of a, b, and we can recover $x = a - b$. Next, notice that a constraint in "greater than or equal" form can be converted to "less than or equal" if we multiply both sides by -1. Finally, an equation can be converted to two "less than or equal"

---

[3]Problem constants are often called *parameters*.
[4]Microsoft Excel

inequalities in a way that may seem artificial:

$$2a + 3b - c = 4 \quad \text{is} \quad \begin{array}{rcr} 2a + 3b - c & \leq & 4 \\ -2a - 3b + c & \leq & -4 \end{array}$$

◇ Problem 19
Express problems (a), (b), and (c) on p.23 in primal form.
◇

We will use a second standard form for a linear program, the *equation form*, in which the constraints are equations.

$$(2.2) \qquad \text{minimize} \quad Z \quad \text{where} \quad Z + C \cdot X = z_0, \quad D \cdot X = B, \quad X \geq \mathbb{O}$$

The constraint $A \cdot X \leq B$ has been replaced by a linear equation $D \cdot X = B$.

Equation form will be necessary as well. As above, we can convert a maximization problem into a minimization problem by considering the negative of the objective, and we can replace unconstrained problem variables so that our problem variables are required to be non-negative. To convert an inequality into an equality, we resort to *slack variables*. For instance, if we have a constraint

$$2 \cdot a - 5 \cdot b \leq 17$$

then we introduce a new (slack) variable s defined by $s = 17 - (2 \cdot a - 5 \cdot b)$. The inequality of the constraint is equivalent to the equation plus the condition $s \geq 0$. Thus,

$$\begin{array}{c} 2 \cdot a - 5 \cdot b \leq 17 \\ a \geq 0, \ b \geq 0 \end{array} \quad \text{becomes} \quad \begin{array}{c} 2 \cdot a - 5 \cdot b + s = 17 \\ a \geq 0, \ b \geq 0, \ s \geq 0 \end{array}$$

For example, if we have constraints in primal form, $A \cdot X \leq B$ with A having size $m \times n$, then we can form an $m \times 1$ column of slack variables S, and

$$A \cdot X \leq B \quad \text{becomes} \quad A \cdot X + S = B$$

◇ Problem 20
Express problems (a) and (b) on p.23 in equation form.
◇

The constraints of a linear program typically have infinitely many feasible vectors. One of the Dantzig's key insights was to see how to confine the search for a solution to the *finite* set of basic vectors.[5]

The proof of the following will fall out later from the simplex algorithm.

PROPOSITION 2.1. *Suppose we have a linear program in equation form: minimize $Z + C \cdot X = z_0$ such that $D \cdot X = B$ and $X \geq \mathbb{O}$. If the linear program is feasible, then there is a basic vector that is a feasible vector. If the linear program has a solution, then there is a solution that is a basic vector.*

Proposition 2.1 reduces both the problem of feasibility and the problem of finding a solution to the problem of checking a finite set of vectors – the set of basic vectors. The proposition leaves open the question of determining whether a feasible linear program has a solution or not. Setting this open question aside, the proposition suggests an algorithm for linear programs: check the value of the objective on the finite set of basic vectors. However, in a problem with a reasonably large number of variables and constraints, there can be a prohibitively large number of basic vectors. The simplex algorithm uses a very clever and efficient search through the set of basic vectors, as we will soon see.

◇ **Problem 21**

How many basic vectors could there be in a system of 50 linear equations in 100 variables,[6] where the coefficient matrix has rank 50? If you could check one trillion basic vectors per second, how long would it take to check them all?

◇

In class we will discuss using a spreadsheet such as Excel[7] to solve LP's numerically. Throughout the present chapter, we will give application problems that should be solved by spreadsheet. These problems will be marked with ★.

---

[5]See chapter 1, section 3 on Linear Equations and Reduced Form.

[6]There are applications of linear programming that involve thousands of variables, and so 100 variables is really not too many!

[7]Microsoft Excel

◇ **Problem 22**

★ Table 1 gives the ash and sulfur content of the coal that comes from three mines (A,B,C). That table also gives the cost per ton of producing (mining) coal from each of the three mines. Find the cheapest way to produce a ton of coal having at most 5% ash and at most 4% sulfur.

Table 1.

	Production cost/ton	Ash %/ton	Sulfur %/ ton
Mine A	$25	5.5	5
Mine B	$27	6	4
Mine C	$30	4	3

◇

◇ **Problem 23**

★ Continuing the previous problem, Table 2 gives the cost of shipping coal from each of the three mines to each of three customers (I,II,III). Customer I gets 25% of each ton of coal produced, Customer II gets 30%, Customer III gets 45%. Determine how much coal is produced at each mine and sent to each customer to minimize the total production and shipping costs. The ash and sulfur limits are still in effect for each customer. It will be convenient to assume that one ton of coal total (all mines, all customers) is produced. How much are we overspending in production costs to minimize the overall costs?

Table 2 – shipping costs $/ton.

	Customer I	Customer II	Customer III
Mine A	20	30	40
Mine B	45	30	35
Mine C	40	60	15

◇

◇ **Problem 24**

★ We produce regular gasoline and premium gasoline from five ingredients: REF (reformate), FCG (cracker gasoline), ISO (isomerate), MTB, and BUT (butane). Each ingredient has three characteristics: RON (research octane number), RVP, ASTM (volatility at 70 degrees Celsius). These characteristics are proportional, so that, for instance, if two gallons of gas has RON 7, and three gallons has RON 6, then the total RON is $2 \cdot 7 + 3 \cdot 6$. There is also a cost per gallon for each ingredient. Here are the characteristics for each of the the five ingredients in units per gallon.

	RON	RVP	ASTM	cost ($ per gal.)
REF	98.9	7.66	-5	1
FCG	93.2	9.78	57	1
ISO	86.1	29.52	107	2
MTB	117	13.45	98	10
BUT	98	166.99	130	1

Determine how 4000 gallons of gasoline, with at least 1000 gallons of premium grade gasoline, should be formulated to minimize the total cost, so that RON and ASTM for each grade are at least the numbers given in the following table, the RVP for each grade is exactly the number in the table, and so that at least 1% of each grade of gasoline is a combination of MTB and ISO. The table numbers are *per gallon*, so if the total RON is R for 1000 gallons of regular gas, then we need $R \geq 1000 \cdot 90$.

	RON	ASTM	RVP	(per gallon)
regular	90	10	21	
premium	96	10	22	

◇

◇ **Problem 25**

★ At the beginning of each week, we have some chairs in storage and we produce additional chairs. It costs $1 to store a chair each week. The costs of production vary: $2 per chair in week 1, $1 in week 2, $3 in week 3. We can produce up to 15 chairs during weeks 1 and 2; 20 during week 3. Chairs are sold from the number in storage at the beginning of the week plus the number produced that week. The maximum we can sell is 13 for week 1, 17 for week 2, and 30 for week 3. We receive $3.50 per chair sold in week 1, $5 per chair in week 2, $7 per chair in week 3. If we start with no chairs in storage, how many chairs should we produce, sell, and store each week to maximize the difference between revenue and costs over three weeks?

2. Canonical Simplex

The simplex algorithm operates on a very special equation form called *canonical form*. Canonical form is a brand of equation form, so here's an equation form: minimize Z such that $Z + C \cdot X = z_0$ and $A \cdot X = B$ and $X \geq \mathbb{0}$, where A is an $m \times n$ matrix, B is $m \times 1$, and C is $1 \times n$. (So there are n variables and m constraints.) This form is *canonical* if

(a) The $m \times n$ coefficient matrix A has rank m and is in reduced form.
(b) If $X[i]$ is a basic variable as indicated by the reduced form, then $C[i] = 0$.
(c) $B \geq \mathbb{0}$.

Let V be the basic vector for the reduced form $A \cdot X = B$, and then the entries of V are 0 at the free variables, and they take on the non-negative entries of B at the basic variables. Thus, $V \geq \mathbb{0}$, and so V is a feasible vector for the linear program. In particular, a linear program in canonical form is necessarily feasible! Furthermore, notice that $C \cdot V = 0$. Indeed, if $X[j]$ is a basic variable, then property (b) shows that $C[j] = 0$. If $X[j]$ is free, then $V[j] = 0$. These two conditions give $C \cdot V = 0$. It follows that $Z(V) = z_0 - C \cdot V = z_0$. We call z_0 the *objective value* of the canonical form.[8]

For both the theory and for (small!) hand calculations, we want to have an augmented matrix form for canonical form. Here's how we will write it. We put the variable names on the top row for reference.

$$(2.3) \qquad \begin{pmatrix} X^T & Z & = \\ A & \mathbb{0} & B \\ C & 1 & z_0 \end{pmatrix}$$

Canonical form can be manipulated into showing whether the linear program has a solution, as we will describe momentarily. You might wonder how we arrive at canonical form for a linear program. We will deal with this in the next section; for now, we mention a special case that occurs quite often. A linear program whose constraints are in primal form $E \cdot X \leq B$, where $B \geq \mathbb{0}$ can be put into canonical form by including slack variables as we did in Section 1. Indeed, if S is a column of non-negative slack variables, then the

---

[8]That z_0 is the objective value is the reason it was included in equation form.

constraints are $E \cdot X + S = B$. In this form, the variables in X are free and the slack variables in S are basic. Write the coefficient matrix with an identity matrix to the right of E; like this: $\begin{pmatrix} E & I_m \end{pmatrix}$. Condition (a) in the definition of canonical form is satisfied, with the slack variables as basic. The objective does not involve the slack variables, and so (b) holds. Since $B \geq \mathbb{O}$, we have condition (c).

The simplex algorithm will perform replacements on canonical form. Eventually, the conditions of the following fact will hold, and we will know whether the linear program has or does not have a solution.

PROPOSITION 2.2. *Assume we have a canonical form* $Z + C \cdot X = z_0$ *and* $A \cdot X = B$.

(a) *If* $C \leq \mathbb{O}$, *then* z_0 *is the minimum of* Z *among all feasible vectors, and the basic vector of* $A \cdot X = B$ *is a solution to the linear program.*

(b) *Suppose that* $C[k] > 0$ *for some* k, *and* $A[i, k] \leq 0$ *for all* i. *Then there are feasible vectors in which* $X[k] \to \infty$ *and* $Z \to -\infty$. *Thus,* Z *has no minimum, and the linear program, although feasible, has no solution.*

PROOF. For (a), suppose that $C \leq \mathbb{O}$. The condition $X \geq \mathbb{O}$ shows that $C \cdot X \leq \mathbb{O}$, and so $-C \cdot X \geq \mathbb{O}$. The equation $Z = z_0 - C \cdot X_2$ then leads to $Z \geq z_0$. This shows that z_0 is a lower bound for Z among all feasible vectors. But $Z = z_0$ at the basic vector, and so the minimum of Z is z_0, and the basic vector is a solution to the LP.

Assume the hypothesis of (b). Notice that $X[k]$ is free, since the basic variables have been eliminated from C. Let $X[k]$ assume any non-negative value, and set the other free variables to 0. If $X[q]$ is the basic variable for the i-th row of $AX = B$, then $X[q] = B[i] - A[i, k]X[k]$ for $1 \leq i \leq m$. We claim that the resulting vector X is feasible.

Indeed, the free variables are all non-negative. As for the basic variable $X[q]$: since $A[i, k] \leq 0$ and $X[k] \geq 0$, we see that $-A[i, k]X[k] \geq 0$, and so for each i with $1 \leq i \leq m$, we have

$$X[q] = B[i] - A[i, k] \cdot X[k] \geq B[i] \geq 0$$

Thus, our specific variables form a feasible vector for every $X[k] \geq 0$. Letting $X[k] \to \infty$, the fact that $C[k] > 0$ and the equation for Z show that $Z \to -\infty$. This shows that Z has no minimum. ∎

When the hypothesis of Proposition 2.2a or Proposition 2.2b holds, we say the canonical form is a *solution form*. Notice that solution form *might indicate* that there is no solution!

Now we will deal with the situation where the hypothesis of Proposition 2.2a,b does not hold. Then there is some $C[k] > 0$ such that there is i with $A[i, k] > 0$. As noted above, the variable $X[k]$ is free; we intend to use it to replace one of the basic variables. We know that we can replace the basic variable in row q provided that $A[q, k] \neq 0$. We describe a particular choice of replacement that will keep us in canonical form and move toward a solution form.

Let F be the set of i such that $A[i, k] > 0$, and we know that F is non-empty. For each $i \in F$, look at the ratio[9] $B[i]/A[i, k]$, and choose i so that this ratio is minimal. We say that i, k is a *feasible replacement*. If $X[j]$ is the basic variable in row i, then we intend to replace $X[j]$ by $X[k]$, so that $X[k]$ will become the basic variable in row i. Shortly we will prove that the system that results from a feasible replacement is canonical.

We will also see that feasible replacements usually decrease the objective value, and so it makes sense that those replacements move the objective toward a minimum. It is possible, but rare, that a feasible replacement leaves the objective value unchanged; we sort this out in the following.

PROPOSITION 2.3. *Assume we have a canonical form $Z + C \cdot X = z_0$ and $A \cdot X = B$. Suppose that q, k is a feasible replacement. Suppose that $X[j]$ is the basic variable for row q of A, and suppose that when $X[j]$ is replaced by $X[k]$, the resulting system is $Z + C' \cdot X = z_0'$ and $A' \cdot X = B'$. Then the new system is canonical. Furthermore, if $B[q] > 0$, then $z_0' < z_0$. If $B[q] = 0$, then $z_0' = z_0$.*

---

[9]This ratio is called a *theta ratio* in many texts. We can describe a feasible replacement by saying that it involves a *minimal theta ratio*.

PROOF. Replacement produces a reduced form for the constraints, and it eliminates new basic variable $X[j]$ from the objective, and so we have conditions (a) and (b) in the definition of canonical form.

To prove (c), let F be the set of rows i such that $A[i, k] > 0$. Replacement produces $B'[q] = B[q]/A[q, k]$ and since $A[q, k] > 0$ and $B[q] \geq 0$, and so (c) holds at row q. For a row $i \neq q$, we see that

$$B'[i] = B[i] - A[i, k]B[q]/A[q, k]$$

If $A[i, k] \leq 0$, then $B'[i] \geq 0$, since $B[i] \geq 0$ and $B[q]/A[q, k] \geq 0$. If $A[i, k] > 0$, then $i \in F$, so that the definition of q, k as feasible is that $B[q]/A[q, k] \leq B[i]/A[i, k]$ and then

$$B'[i] = A[i, k] \cdot \Big(B[i]/A[i, k] - B[q]/A[q, k] \Big)$$

and this is non-negative. Thus, (c) holds.

We see that $z_0' = z_0 - C[k]B[q]/A[q, k]$. If $B[q] = 0$, we have $z_0' = z_0$, if $B[q] > 0$, then since $C[k] > 0$ and $A[q, k] > 0$, we have $z_0' < z_0$. ∎

This last proposition holds the key to our main theorem on linear programs: Dantzig's Theorem. We will see that if we start with canonical form, there will always be a sequence of feasible replacements that puts us in solution form. In other words, we will keep applying Proposition 2.3 until one of the conditions of Proposition 2.2 holds.

◇ Problem 26
Here is the augmented matrix of a linear program in canonical form. Find all feasible choices for replacement. Perform them and check that the conclusions of Proposition 2.3 hold.

x_1	x_2	x_3	x_4	x_5	x_6	x_7	Z	$=$
1	1	0	−1	0	0	0	0	2
0	−1	1	2	2	0	0	0	1
0	2	0	1	6	1	0	0	4
0	0	0	3	−3	0	1	0	0
0	3	0	−2	4	0	0	1	3

Suppose we have an LP in canonical form with constraints $AX = B$. It is not hard to see that a choice of basic variables for this equation determines the objective value at the basic vector for this choice, for the choice of basic variables determines the free variables, and we get the basic vector by setting the free variables to 0, from which the basic variables are determined. This leads to exactly one basic vector V, and one objective value $Z(V)$.

Proposition 2.3 shows that the objective value of a canonical form decreases whenever a feasible replacement is made with a positive right side $B[q]$. Therefore, if a sequence of such replacements is made, we will never encounter the same set of basic variables more than once – there is no way to increase the objective value back to a previously encountered value. There are only finitely many sets of basic variables, and so we can undergo replacements with positive right side only finitely many times: eventually we will run out of feasible replacements (i.e. we reach solution form!), or we will reach a form in which the only feasible replacements involve 0 right sides.

In practice, the latter possibility is rare – usually there are feasible replacements with positive right side. If the right sides are positive under every sequence of feasible replacements, then the LP is called *non-degenerate*. We will not pursue this condition, except to say that it is rather common. The point is that if an LP is non-degenerate, then it is obvious that finitely many feasible replacements will always lead to solution form.

But it is possible that an LP is *degenerate* and we may be forced to consider 0 right sides. We can still prove that there is a sequence of feasible replacements that lead to solution form, and we will undertake that argument in the next subsection. When we work problems by hand we will simply trust our choices to make progress to solution form, even if 0 right sides occur. When we use Excel, we will trust that program's ability to find a solution. Thus, our concern with 0 right sides is rather theoretical. We use the name *canonical simplex algorithm* for the application of feasible replacements to bring a canonical LP into solution form.

◇ **Problem 27**

Use the canonical simplex algorithm by hand to solve the following.

(a) Maximize $x_1 - 2x_2 + 3x_3$ such that $x_1 - 2x_2 + x_3 \leq 10$, $x_1 - x_2 + x_3 \leq 10$, $x_1 \geq 0$, $x_2 \geq 0$, and $x_3 \geq 0$.

(b) Minimize $-x$ subject to $3x + y + v \leq 10$ and $5x - y \leq 15$ and x, y, v are non-negative.

◇

2.1. Special Canonical Form. This subsection describes a foolproof way of obtaining solution form, even if 0 right sides occur. You can skip reading this section, if you wish, although it will be quoted in the section on Feasibility.

What we need is a way to put vectors in order. We will describe feasible replacements that make the objective coefficients (C, z_0) decrease with respect to this order. This will have an effect similar to the case of positive right sides: once a set of basic variables occurs, it cannot occur again.

For $u, v \in \mathbb{R}^k$ with $u \neq v$, let i be the maximal subscript where u, v disagree. Thus, $u[i] \neq v[i]$, but $u[j] = v[j]$ for all $j > i$. We say that $u \succ v$ if $u[i] > v[i]$ and we say $v \succ u$ if $v[i] > u[i]$. Obviously either $u \succ v$ or $v \succ u$, but remember that we are assuming the $u \neq v$. We call the ordering thus obtained *lex-top ordering*.

◇ **Problem 28**

If $u, v, w \in \mathbb{R}^k$ and $u \succ v$ and $v \succ w$, then $u \succ w$. Let $u, v \in \mathbb{R}^k$ with $u \neq v$. Show that $u \succ v$ if and only if $u - v \succ \mathbb{O}$.

◇

To set up our use of lex-top ordering, we need some notation. We are given an LP in equation form: minimize Z such that $A \cdot X = B$ and $Z + C \cdot X = z_0$ and $X \geq \mathbb{O}$. We will write R_i for the i-th row of the augmented matrix $[A|B]$. We suppose that A is $m \times n$ and in reduced form with rank m. We will say that the LP is in *special canonical form* if $R_i \succ \mathbb{O}$ for all i. It follows that $B \geq \mathbb{O}$, for if $B[i] < 0$ for some i, then we see that $R_i \not\succ \mathbb{O}$.

Special canonical form can always be obtained from the equation form by switching columns, if necessary, so that the basic variables are as far to the right as possible. Then if $B[i] > 0$, we see that $R_i \succ \mathbb{O}$. If $B[i] = 0$, then, reading R_i right to left, the first non-zero entry we come to is the pivot 1 for the basic variable of row i. Thus, $R_i \succ \mathbb{O}$, and we see that we are in special canonical form. If we switch columns to obtain special canonical form, then, in applying the algorithm we are about to describe, we need to keep the same ordering of columns throughout, so that the specific lex-top ordering does not change.

PROPOSITION 2.4. *Supposing we have an LP in special canonical form, with the notation just given, suppose that $A[q, k] > 0$ for some q, k, and suppose that for every j with $A[j, k] > 0$, we have $R_j/A[j, k] \succ R_q/A[q, k]$. Then q, k is a feasible replacement, and if $C[j] > 0$, then the replacement changes (C, z_0) into (C', z_0') with $(C, z_0) \succ (C', z_0')$. After replacement, we are still in special canonical form.*

PROOF. Since $A[q, k] > 0$, we see that $R_q/A[q, k] \succ \mathbb{O}$. The replaced form of row q is $R_q' = R_q/A[q, k]$.

We see that $(C', z_0') = (C, z_0) - C[j] \cdot R_q'$. If $C[j] > 0$, then $C[j] \cdot R_q' \succ \mathbb{O}$, and it follows that $(C, z_0) \succ (C', z_0')$.

For a row $i \neq q$, replacement changes R_i into R_i' where

$$R_i' = R_i - A[i, k] \cdot R_q'$$

If $A[i, k] \leq 0$, then since $R_i \succ \mathbb{O}$ and $R_q' \succ \mathbb{O}$, we see that $R_i' \succ \mathbb{O}$. If $A[i, k] > 0$, then by hypothesis, we have

$$R_i/A[i, k] \succ R_q/A[q, k] \quad \text{so that} \quad R_i \succ A[i, k] \cdot R_q'$$

and it follows that $R_i' \succ \mathbb{O}$.   ■

A replacement made as in the hypothesis of Proposition 2.4 is a *lex-top replacement*. Proposition 2.4 proves that a lex-top replacement is a feasible replacement. Furthermore, the proposition shows that the system of equations that results from replacement is still in special canonical form.

Now we imagine that we have a special canonical form with augmented constraint matrix $[A|B]$ all of whose rows R_i satisfy $R_i \succ \mathbb{O}$. If the LP is not in solution form, we can choose k so that $C[k] > 0$ and there is i such that $A[i, k] > 0$. Let F be the set of i such that $A[i, k] > 0$, as before. Because the rows of $[A|B]$ are independent, the vectors $R_i/A[i, k]$ (for $i \in F$) are distinct for distinct i. Thus, there is one that is minimal: there is $q \in F$ such that if $j \in F$ and $j \neq q$, then $R_j/A[j, k] \succ R_q/A[q, k]$. Proposition 2.4 says that q, k is a feasible replacement, that the rows R_i' of the resulting canonical form satisfy $R_i' \succ \mathbb{O}$ and that the resulting objective (C', z_0') satisfies $(C, z_0) \succ (C', z_0')$.

We can continue. As we perform *lex-top replacements* the objective vectors (C, z_0) *descend* with respect to the ordering. As we argued in the canonical case, this implies that a given set of basic variables can occur at most once. Sooner or later, we must obtain solution form.

We have just proved that there is always a sequence of feasible replacements that change a canonical form into solution form. We thereby obtain one of the claims of Proposition 2.1: if a linear program in canonical form has a solution, then it has a solution that is a basic vector, for the canonical simplex algorithm produces a solution form, and in that form there is a basic vector that produces the minimum value of the objective.

We remind you what we said before: In practice, we will not worry about lex top ordering at all – calculations done by hand will never cause trouble and those done in Excel will depend on that program's ability to move to a solution.

3. Feasibility

Given an arbitrary LP, if we can put it into canonical form, then we can solve it by the canonical simplex algorithm. Canonical form is necessarily feasible, so if an LP can be put into canonical form, then the LP is feasible. It turns out that the converse is true: if an LP is feasible, then it can be put into canonical form. We now describe how to do this; it will involve an algorithm very similar to the canonical simplex algorithm in that it consists of feasible replacements.

We begin by writing the constraints in equality form. Thus, slack variables need to be included in any constraints that are inequalities. If there are constraints that are equations, we use elimination to put all the equations into reduced form. If the reduced form shows that the constraints are inconsistent, then the LP is infeasible, and we are done. Otherwise, the reduced form expresses basic variables in terms of free variables. (We can discard any "$0 = 0$" equations that result.) We clear these basic variables from the inequality constraints – each inequality constraint already has a basic variable in the form of a slack variable.

We write out reduced form with *non-negative right sides*: if a right side is negative, for instance as in $x - 3 \cdot y = -4$ with basic variable x, then we change to $-x + 3 \cdot y = 4$. In the resulting equation, the coefficient 1 on the basic variable x has become -1. Such equations can come about in two ways. If we have an inequality such as $x + y \geq 5$, then the slack variable s needs to be *subtracted* to keep the right side positive: $x + y - s = 5$. The basic variable s has a -1. Second, an equation obtained from reducing the set of equation constraints might have a negative right side, and we multiply by -1, as just shown.

The reduced form of the constraints can then be separated into two parts: those equations with basic coefficient $+1$ and those equations with basic coefficient -1. The augmented matrix of the form will look like this, with variables X, S_1, S_2, where S_1, S_2 are basic variables.

(2.4)
$$
\begin{array}{cccc}
X & S_1 & S_2 & = \\
A_1 & I & \mathbb{O} & B_1 \\
A_2 & \mathbb{O} & -I & B_2
\end{array}
$$

where the occurrences of I denote identity matrices of the correct sizes. Also $B_1 \geq \mathbb{O}$ and $B_2 > \mathbb{O}$. If there are rows with negative basic coefficients, we do what is called *Phase 1*; otherwise, we skip down to Phase 2, described after the next subsection.

3.1. Phase 1. We append a row to the bottom of the augmented form: each entry in this row is the sum of entries above it in the rows corresponding to A_2. This row is called the *feasible row*. The form (2.4) now looks like this:

(2.5)

$$
\begin{array}{ccccc}
X & S_1 & S_2 & = \\
A_1 & I & \mathbb{O} & B_1 \\
A_2 & \mathbb{O} & -I & B_2 \\
F & \mathbb{O} & -J & f
\end{array}
$$

Thus, $F[j]$ is the sum of column j of A_2, and $-J$ denotes a row of -1's that come from the $-I$ above it, and f is the sum of the entries in B_2.

The position of the entry called f is the *feasible sum*. We treat the matrix (2.5) as if it were a canonical form. We apply feasible replacements, treating the bottom row as the objective. If the row chosen for replacement has a negative basic coefficient, then the replacement produces a basic coefficient 1 in that row. In this case the set of rows with negative basic coefficients loses one member. On the other hand, if the row chosen for replacement has a positive basic coefficient, then the set of rows with negative basic coefficients stays the same. It is straightforward to show, in either case, that the feasible row retains its identity as the sum of the rows with negative basic coefficients. Because of this, we can never encounter the solution form of Proposition 2.2b, for if there is a positive entry in the feasible row, then since that entry is the sum of the entries above it in its column, one of those entries above it must be positive.

Proposition 2.3 is also in force. Feasible replacements decrease or leave the same the feasible sum, called f in (2.5). As is the case with canonical form, there is a sequence of feasible replacements that lead us to solution form. As with canonical form, 0 right sides are possible, and we do not worry about them – we can apply the technique of the subsection on Special Canonical Form if we want to be absolutely certain.[10]

Once we reach solution form, we must be in the situation of Proposition 2.2a. Write the feasible row in the following form:

$$G_0 \cdot X + G_1 \cdot S_1 + G_2 \cdot S_2 = h$$

---

[10]Notice that the rows corresponding to A_2 have positive right sides, and so they are positive in lex-top ordering.

where G_0, G_1, G_2 are non-positive. We know that h is the sum of the right sides coming from rows with negative basic coefficients. Thus $h \geq 0$. If $h > 0$, then we claim that the LP is infeasible. Indeed, if $X, S_1, S_2 \geq \mathbb{O}$, then

$$G_0 \cdot X + G_1 \cdot S_1 + G_2 \cdot S_2 \leq 0 < h$$

and so the feasible row equation cannot hold. Because that equation is the sum of certain of the constraint equations, it would have to hold at a feasible vector. Thus, the LP is infeasible.

If $h = 0$, then each of the non-negative right sides corresponding to a negative basic coefficient must be 0. Each of these rows can be multiplied by -1, and this gives us a canonical form for the LP, once we include the objective equation and eliminate basic variables from it.

3.2. Phase 2. Now we have a canonical form for the LP, either because it fell out of Phase 1 or because we didn't need to do Phase 1. We apply the Canonical Simplex Algorithm to the canonical form to obtain solution form. This is Phase 2.

When Phases 1 and 2 are put together, the entire procedure is the *Simplex Algorithm*. We have put an algorithm summary on p.42.

It is of obvious interest to return to all the linear programs mentioned so far and obtain solutions. Here are some additional problems.

◇ **Problem 29**

Solve these problems *by hand*.

(a) Minimize Z such that $Z - 3x + 5y = 0$, $x - 2y = 3$, $3x + y \leq 20$, $x, y \geq 0$.

(b) Maximize Z such that $Z = 2x + 6y - 2$, $x + 4y \leq 10$, $y - 3x \geq -3$, $x \geq 0$, y arbitrary.

(c) Minimize $2x + 3y$ such that $-x + y \geq 3$, $y - 5x \leq 2$, $x, y \geq 0$.

◇

◇ **Problem 30**

Solve these linear programs using the simplex algorithm *by hand*.

(a) Among all points in the xy-plane on or inside the triangle with vertices $(2,9)$, $(5,3)$, $(7,4)$, maximize $2x + 3y$. (Hint: equations for the boundary lines give inequalities.)

(b) Minimize $2a - 6b - c$ subject to $a - 2b - c \geq 2$ and $2a - 3b \leq 6$, and $a \geq 0$, $b \geq 0$, $c \geq 0$.

◇

+++ The Simplex Algorithm +++

Set-up

Step 0.1. Add slack variables to each inequality constraint as basic variables.

Step 0.2. Put the equation constraints in reduced form, eliminating basic variables from all constraints and from the objective equation.

Step 0.3. If the system is inconsistent, STOP – the LP is infeasible. Make all the constraint right sides non-negative. If there are negative basic coefficients, go to Phase 1, otherwise go to Phase 2.

Phase 1

Step 1.1. Form the feasible row: the sum of the rows with negative basic coefficients.

Step 1.2. If every entry in the feasible row other than the feasible sum is non-positive, go to Step 1.4.

Step 1.3. Find a positive entry in the feasible row; find and make a feasible replacement in that column. Eliminate also in the objective equation. Go to Step 1.2.

Step 1.4. If the feasible sum is positive, STOP – the LP is infeasible. Multiply rows with negative basic coefficients by -1, and go to Phase 2.

Phase 2 – Canonical Form

Step 2.1. If every entry in the objective row other than the objective value is non-positive, STOP –we have a basic solution and the objective value is the minimum of the objective.

Step 2.2. Choose an entry in the objective row that is positive. If the coefficients in the column above that entry are non-positive, STOP – the objective can go to $-\infty$, and so there is no solution. Otherwise, choose and perform a feasible replacement. Go to Step 2.1.

4. The Linear Algebra of Primal Form

For the sections following this one, we need some equations that relate the parts of an LP in primal form to the parts of that same LP at some point in the Simplex Algorithm. Let's start with an LP in primal form: minimize

Z such that $Z + C \cdot X = z_0$ and $A \cdot X \leq B$ and $X \geq \mathbb{O}$, where X holds the problem variables, and where A is $m \times n$. We are *not* assuming that the entries of B are non-negative. To apply the Simplex Algorithm, we include slack variables to obtain equation form: let S be an $m \times 1$ column of slack variables, and write $A \cdot X + S = B$ for the constraints. Here is the augmented matrix of the LP, writing the variable names on the top row.

$$(2.6) \qquad \begin{pmatrix} X^T & S^T & Z & = \\ A & I_m & \mathbb{O}_{m \times 1} & B \\ C & \mathbb{O}_{1 \times m} & 1 & z_0 \end{pmatrix}$$

Now suppose that elementary operations are applied to this system. As in the Simplex Algorithm, the objective variable Z needs to stay basic in this system. Suppose that the following system results:

$$(2.7) \qquad \begin{pmatrix} X^T & S^T & Z & = \\ L & M & \mathbb{O}_{m \times 1} & N \\ D & E & 1 & z_1 \end{pmatrix}$$

We have written the parts of this matrix to correspond directly to the parts in (2.6). Notice that the 0's in the Z-column persist, since Z never loses its status as a basic variable.

We know that the new system (2.7) has exactly the same solutions as the original system (2.6). For instance, $X = \mathbb{O}$ and $S = B$ and $Z = z_0$ is a solution to the original equation, and we conclude that

$$M \cdot B = N \quad \text{and} \quad E \cdot B + z_0 = z_1$$

Furthermore, for every $n \times 1$ matrix V, we get a solution to the original equation: $X = V$, $S = B - A \cdot V$ and $Z = z_0 - C \cdot V$. This is a solution to the new equation; we are interested in the equation for Z:

$$z_0 - C \cdot V + D \cdot V + E \cdot (B - A \cdot V) = z_1$$

Since $E \cdot B + z_0 = z_1$ we see that

$$D \cdot V = (E \cdot A + C) \cdot V$$

This equation is true *for all* $n \times 1$ matrices V, and it follows that $D = E \cdot A + C$.

We record these equations.

PROPOSITION 2.5. *Given the augmented form (2.6) of a primal LP, with problem variables X and slack variables S, suppose that when some sequence of elementary row operations are applied, (2.7) gives the augmented representation of a result. Then we have the following equations.*

$$(2.8) \qquad\qquad E \cdot A + C = D, \quad M \cdot B = N, \quad E \cdot B + z_0 = z_1$$

∎

We will use this proposition quite generally in the next section. Our immediate purpose is to think about the case where the primal LP has a solution, and the form (2.7) is a *solution form*. In that case the objective coefficients – the matrices D, E – constitute what will be called the *Lagrange Multiplier*. The equations that D, E satisfy in (2.8) give a special case of what will be called the *Kuhn-Tucker Conditions* in the context of arbitrary optimization problems. We collect the linear program version of Kuhn-Tucker in the following result. At this point, the conditions will not be memorable; later, when we have the Kuhn-Tucker Conditions in general, we will see that the conditions here are a special case.

PROPOSITION 2.6. *Let A be $m \times n$ and B be $m \times 1$ and C be $1 \times n$ and $z_0 \in \mathbb{R}$. Consider the primal LP: minimize Z such that $Z + C \cdot X = z_0$ and $A \cdot X \le B$ and $X \ge \mathbb{O}$. This LP has a solution $X = V$ if and only if $V \ge \mathbb{O}$ and $AV \le B$ and there is a $1 \times m$ matrix E and a $1 \times n$ matrix D such that*

(a) $E \le \mathbb{O}$ and $D \le \mathbb{O}$.
(b) $-C = E \cdot A - D$
(c) $E \cdot (B - A \cdot V) = 0$ and $D \cdot V = 0$.

PROOF. Write $AX \le B$ as $AX + S = B$ for non-negative slack variables S, as before.

Suppose that V is a solution to the LP. We are not assuming that V is basic, but since it is a solution, the simplex algorithm can be used to find a solution form $Z + D \cdot X + E \cdot S = z_1$ for the objective. In solution form (a) holds, and the equations in Proposition 2.5 hold, so that (b) holds as well.

The solution form shows that z_1 is the minimum value of Z. Thus, $Z(V) = z_1$.

Let $U = B - A \cdot V$ and we see that $U \geq \mathbb{O}$ has to give the values of the slack variables. Thus, $Z + D \cdot V + E \cdot U = z_1$. We know that $Z = z_1$ here, so we have $D \cdot V = -E \cdot U$. Since $E \leq \mathbb{O}$ and $U \geq \mathbb{O}$, we see that $E \cdot U \leq 0$, and so $D \cdot V \geq 0$. On the other hand, $D \leq \mathbb{O}$ and $V \geq \mathbb{O}$, so that $D \cdot V \leq 0$, and now we have $D \cdot V = 0$. It follows that $E \cdot U = 0$, and this equation is $E \cdot (B - A \cdot V) = 0$. We have (c).

Conversely, suppose that V, D, E satisfy the conditions given. (For this direction, we cannot assume that we have a solution form, so Proposition 2.5 is not necessarily relevant!) We claim that V is a solution to the LP. Use (b) and (c) to compute

$$Z(V) = z_0 - C \cdot V = z_0 + (E \cdot A - D) \cdot V = z_0 + E \cdot A \cdot V = z_0 + E \cdot B$$

On the other hand, if Y is a feasible vector, then (b) yields

$$Z(Y) = z_0 - C \cdot Y = z_0 + E \cdot B - E \cdot (B - A \cdot Y) - D \cdot Y$$

Since $E \leq \mathbb{O}$ and $B - A \cdot Y \geq \mathbb{O}$ and $D \leq \mathbb{O}$ and $Y \geq \mathbb{O}$, we see that $Z(Y) \geq Z(V)$, so that $X = V$ solves the LP. ∎

5. Duality

It is a profound fact that the simplex method actually solves two problems at the same time: the stated minimization problem and a "dual" maximization problem. To explain this, we first give a very applied example. The arithmetic involved in this problem is very simple; the point of the problem is to study the problem coefficients from two different points of view.

Example. We have up to \$10 to buy x loaves of bread and y rounds of cheese. Bread costs \$2 per loaf; cheese costs \$3 per round. We can carry 5 pounds total. A loaf of bread weighs 2/3 pounds; a round of cheese weighs 2 pounds. We want to purchase bread and cheese to maximize the total calories: bread has 500 calories per loaf; cheese has 1000 calories per round. ∎

We can use the Simplex Algorithm to solve this problem: the maximum is 8750/3, and it occurs when $x = 5/2$ and $y = 5/3$. The way the problem is stated, calories come from bread and cheese. Because money and weight are used to determine how much bread and cheese we obtain, we might try to explain the calories as a function of money and weight. We imagine a variable u that estimates calories per dollar, and a variable v that estimates calories per pound. Bread costs \$2 per loaf, so $2 \cdot u$ estimates calories per loaf coming from cost. Bread weighs 2/3 pounds per loaf, so $(2/3) \cdot v$ estimates calories per loaf coming from weight. The expression $2 \cdot u + (2/3) \cdot v$ attempts to explain total calories per loaf. There may be some inefficiency in this explanation – there may be some overlap in the calories coming from the cost and from the weight. Thus, we write

$$2 \cdot u + (2/3) \cdot v \geq 500$$

to indicate that our explanation needs to cover the actual number of calories in a loaf. Similarly,

$$3(\text{dollars/round}){\cdot}u(\text{calories/dollar})$$
$$+ 2(\text{pounds/round}) \cdot v(\text{calories/pound})$$
$$\geq 1000(\text{calories/round})$$

says that u, v cover the number of calories in a round of cheese. We can spend up to \$10, and so $10 \cdot u$ is the maximum number of calories we can purchase. We can carry up to 5 pounds, and so $5 \cdot v$ is the maximum number of calories we can carry. We are regarding cost and weight as separate, and so it is natural to consider the sum $W = 10 \cdot u + 5 \cdot v$. This quantity is related to the original objective $Z = 500 \cdot x + 1000 \cdot y$ in the following way:

$$Z = 500 \cdot x + 1000 \cdot y \leq 10 \cdot u + 5 \cdot v = W$$

whenever x, y, u, v satisfy their constraints.[11] In other words, to decide whether W can explain the solution to the original problem (maximizing Z), we would *minimize* W. We mentioned that there may be some inefficiency in the way

---

[11]We will derive this inequality very generally in a later proposition.

u, v overlap; minimizing W attempts to wash out that inefficiency as much as possible.

Here are the two problems – the original problem and the new problem.

$$\max \quad 500x + 1000y$$
$$\text{where} \quad 2x + 3y \leq 10,\ (2/3)x + 2y \leq 5,\ x, y \geq 0$$
$$\min \quad 10u + 5v$$
$$\text{where} \quad 2u + (2/3)v \geq 500,\ 3u + 2v \geq 1000,\ u, v \geq 0$$

This pair of LP's gives an example of *dual LP's*.

To describe the dual in general, we revisit the primal form of a linear program:

(2.9) $\qquad$ min. Z with $AX \leq B$ and $Z + CX = z_0$ and $X \geq 0$

Let the size of A be $m \times n$. The *dual program* has a $1 \times m$ matrix Y of variables:

(2.10) $\qquad$ max. Z' such that $YA \geq C$ and $Z' + YB = z_0$ and $Y \geq \mathbb{0}$

Observe that the inequality $YA \geq C$ involves n individual inequalities. Thus, if the primal problem has m constraints and n variables, the dual has n constraints and m variables. Furthermore, in the product YA, the variable $Y[i]$ is multiplied by entries in the i-th row of A, and this row corresponds to the i-th constraint. We think of dual variable $Y[i]$ as belonging to the i-th constraint.

◇ Problem 31
Show carefully that the two bread and cheese problems are dual to each other.

We want to show that if the dual (2.10) is regarded as primal, then the primal becomes the dual. In other words, the dual of the dual is the original problem! To see this, we need to write (2.10) in primal form, with the variables as a column rather than as a row and with the constraints in "less than or equal to" form. The transpose accomplishes this; recall that the transpose reverses multiplication; if we take the transpose of all the equations in (2.10), we get the following (messy!) equivalent problem:

$$\text{min.} \; -Z' \text{ with } -A^T Y^T \leq -C^T, \; -Z' - B^T Y^T = -z_0, \; Y^T \geq \mathbb{O}$$

In class, we will show that it follows that the dual of the dual is the original problem.

We remark that there is more than one way that a given linear program can be put into primal form. For instance, if we include vacuous constraints, we add to the number of *variables* in the dual. Many texts refer to "the" dual of a linear program, as if dual programs were unique; this over-precision is usually harmless. The various dual programs are related to each other, but in order to obtain the important facts quickly, we will pursue this to some extent in problems.

◇ Problem 32
Determine a dual linear program to each of the following.

(a) Minimize $3 - 5a - 2b - 2c$ such that $a + 3b + c \leq 10$ and $a + b + c \geq 2$ and $a \geq 0$, $b \geq 0$, $c \geq 0$.

(b) Define $f(a, b)$ to be the maximum of $2a + b$ and $3a - 4b$. Find the minimum of f such that $a \geq 0$, and $b \geq 0$.

◇ Problem 33
Consider this problem: minimize $Z = C \cdot X$ such that $A \cdot X \leq B$. (There is no sign restriction on the problem variables X.) Show that the dual problem can be stated like this: minimize $Y \cdot B$ such that $Y \cdot A + C = \mathbb{O}$ and $Y \geq \mathbb{O}$.

◇

◇ Problem 34
Consider this problem: minimize $Z = C \cdot X$ such that $A \cdot X = B$ and $X \geq \mathbb{O}$. Show that the dual problem can be stated like this: mimimize $Y \cdot B$ such that $Y \cdot A \leq C$.

Recall that if U is a feasible vector for (2.9), then $Z(U) = z_0 - C \cdot U$ is the value of the objective at that vector. In the bread and cheese problem, we mentioned an inequality between the objectives of the primal and dual; here is that inequality in general.

SEPARATION THEOREM. *Given the primal linear program (2.9) and its dual (2.10), let U be a feasible vector for (2.9) and let V be feasible for (2.10). Then $Z(U) \geq Z'(V)$.*

PROOF. Direct calculation. If we multiply the inequality $AU \leq B$ by V, then because $V \geq 0$, the inequality still holds, and we get $VAU \leq VB$. Since $U \geq 0$, we can multiply $VA \geq C$ by U and we get $VAU \geq CU$. Putting $VAU \leq VB$ and $VAU \geq CU$ together, we get $CU \leq VB$. Bringing in the equations in Z and Z', we have $z_0 - Z(U) \leq z_0 - Z'(V)$, which is what we wanted.   ∎

COROLLARY 2.7. *If a primal linear program is feasible but has no solution, then its dual program is infeasible. If the dual program is feasible but has no solution, then the primal program is infeasible.*

PROOF. In the first case, we allow $Z \to -\infty$ with various X, feasible for the primal program. If the dual program were feasible, the Separation Theorem would give a value of Z' as a lower bound for all the Z-values. Thus, the dual program cannot be feasible.

The second statement is proved similarly, since $Z' \to \infty$ in that case.   ∎

Now we can prove the important theorem that a primal linear program has a solution if and only if the dual program has a solution. Furthermore, the extreme values of the objectives of each program are the same! This is called the *Duality Theorem*. This theorem has an interesting history; see [1].

Notice that the proof explains how to find a solution to the dual program from the solution form of the primal.

DUALITY THEOREM. *The primal linear program (2.9) has a solution if and only if the dual linear program (2.10) has a solution. The minimum of Z is equal to the maximum of Z'.*

PROOF. The primal form can be converted to equation form by including slack variables, as was done in (2.6) on p.43. We assume that the solution form is in (2.7), and so the equations of Proposition 2.5 hold. We will show that $-E$ is a solution to the dual program. Observe that E is $1 \times m$, and so it has the right size. Since (2.7) is a solution form, we have $E \leq \mathbb{O}$ so that $-E \geq \mathbb{O}$. For the same reason $D \leq \mathbb{O}$, and so the equation $D - E \cdot A = C$ shows that $-E \cdot A \geq C$. Thus, $-E$ satisfies the constraints of the dual. For the dual objective Z', we have $Z' = z_0 + E \cdot B = z_1$, where z_1 is a value of the objective Z of the primal program. The Separation Theorem shows that z_1 is an upper bound for Z'; we see that z_1 is the maximum for Z'. Thus, $-E$ solves the dual program.

Conversely, since the primal program is the dual of the dual, if the dual program has a solution, then the primal has a solution with the same objective.

Notice how to find a solution for the dual: let $Y[i]$ be the dual variable that goes with the i-th constraint, and suppose that $S[i]$ is the slack variable for that constraint in the original primal form. Let $E[i]$ be the coefficient of $S[i]$ in the *solution form of the primal problem*. Then $Y[i] = -E[i]$ is the value of $Y[i]$ at a solution to the dual linear program.

◇ Problem 35

Solve each of the LP's (a) and (b) on p.48 by hand, and read the solution to the dual problem from the solution form of the primal program. Solve one of the dual problems as if it were primal and read a solution to the original problem.
◇

Given dual linear programs, we have shown that if the objective of one is unbounded, the other program is infeasible. The Duality Theorem shows that if one has a solution, so does the other. We might wonder whether it is possible for both programs to be infeasible. Here is an example of an infeasible program whose dual is also infeasible.

$$\text{min. } Z \text{ such that } x_1 - x_2 \leq 5, \quad -x_1 + x_2 \leq -7, \quad Z + 3x_1 - x_2 = 0$$

There is an orthogonality condition that surfaces in the proof of the Duality Theorem. We claim that Y and S can be chosen in the solution of the dual and primal programs, respectively, so that $Y[j] \cdot S[j] = 0$ for each j (for each constraint of the primal program). Among other things, this implies that the dot product of Y and S is 0, so that they are orthogonal. To see that $Y[j] \cdot S[j] = 0$, first notice that this is trivial if $Y[j] = 0$. To consider the case $Y[j] \neq 0$, look at the proof of the Duality Theorem, where the solution $Y = -E$ to the dual problem is used. If $Y[j] \neq 0$, then $E[j] \neq 0$, and remember that $E[j]$ is the coefficient of $S[j]$ in the objective Z. The simplex algorithm that produces E gives non-zero coefficients only at free variables. When the basic vector to the primal program is used, the free variables are all 0. Thus, $S[j] = 0$ in the basic vector, and we have $Y[j] \cdot S[j] = 0$ in this case as well. The condition that $Y \cdot S = 0$ goes by the name of *complementary slackness* since the Y's have non-zero coefficients where the S's (the slack variables) are non-zero and vice-versa.

The Duality Theorem has many equivalent forms. Here is a typical use that has relevance to the convex problems in Chapter 6.

PROPOSITION 2.8. *Let n be a positive integer, and let $u, v_1, v_2, \ldots, v_m \in \mathbb{R}^n$. Suppose that for all $w \in \mathbb{R}^n$ such that $w \circ v_i \geq 0$ for all i, we necessarily have $w \circ u \geq 0$. Then u is a non-negative linear combination of the v_i.*

PROOF. Write u and the v_i as $n \times 1$ matrices. Let A be the $n \times m$ matrix whose columns are the v_i.

Consider the LP: maximize W such that $W + y \cdot u = 0$ and $y \cdot A \geq \mathbb{O}$. The variables are the $1 \times n$ matrices y, unconstrained as to sign. This problem is feasible, for $y = \mathbb{O}$ is a feasible vector. Furthermore, if $y \cdot A \geq \mathbb{O}$, then $y \circ v_i \geq 0$ for each i, and so $y \circ u = y \cdot u \geq 0$. It follows that $W \leq 0$ for all feasible vectors. Since W is bounded above, the feasible LP has a solution. By the Duality Theorem, the dual problem has a solution; that problem is to mimimize Z such that $Z + \mathbb{O} \cdot X = 0$ and $A \cdot X = u$ and $X \geq \mathbb{O}$. The entries of a feasible vector X show how to write u as a non-negative combination of the v_i.   ∎

6. Perturbation and Shadow Prices

Going back to the bread and cheese problem on p.45, we said we could spend up to \$10. The number 10 represents the amount of a resource available. It is typical to regard the right side of the constraints in an LP as *resource bounds* – available amounts of some material. We want to know what happens to the problem solution when we change 10 slightly. To understand this situation, we need to distinguish between the objective Z of the problem and the *minimum* objective, which we now call $\bar{Z}$. The objective Z is a function of the problem variables, with the resource bound constant. We will see that the minimum objective $\bar{Z}$ is a function of the resource bound, not involving the problem variables.

Let's change the bound from 10 to $10 + \delta$, where δ is a new variable, measuring the *perturbation*[12] of 10. We will re-work the bread and cheese problem with the perturbed right side; here is the initial form and solution form. The variable s is slack for the first constraint, and t for the second. We made the same replacements as for the non-perturbed case $\delta = 0$.

$$\begin{pmatrix} x & y & s & t & = \\ 2 & 3 & 1 & 0 & 10+\delta \\ 2/3 & 2 & 0 & 1 & 5 \\ 500 & 1000 & 0 & 0 & 0 \end{pmatrix} \longrightarrow \begin{pmatrix} x & y & s & t & = \\ 1 & 0 & 1 & -3/2 & 5/2+\delta \\ 0 & 1 & -1/3 & 1 & 5/3-\delta/3 \\ 0 & 0 & -5/3 & -5/2 & -8750/3-5\delta/3 \end{pmatrix}$$

We notice several suggestive features: First, notice that the final form is still a solution form, since the coefficients $-5/3$ and $-5/2$ are the same as in the unperturbed problem. Indeed, since δ starts out only in the right side column, δ stays only in that column in the final form. Second, the coefficients of δ in the final form are the coefficients in the column of s, and s is the

---

[12]The word *perturbation* means *small variation*.

slack variable for the perturbed constraint. Third, the minimal objective is $\bar{Z} = -(8750 + 5 \cdot \delta)/3$, a function of δ, as we predicted. The change in $\bar{Z}$ with respect to δ is $d\bar{Z}/d\delta = -5/3$; that derivative is called the *shadow price*. The word *price* is used since price is the derivative with respect to a (purchased) resource. Our second observation identifies the shadow price with the objective coefficient in solution form of the slack variable s. Fourth, the final form is canonical **if** the right side coefficients are non-negative – we can compute that this happens exactly when $-5/2 \leq \delta \leq 5$. The inequality on δ defines what we call an *allowable perturbation* – one that allows the final form to remain in canonical solution form.

We will now show that these observations generalize. The solution form of a primal LP tells us two things: which perturbations of the constraint are allowable, and what is the shadow price of the constraint. To set this up, assume we have a primal LP, expressed as in (2.6), using slack variables S. Suppose that the coefficient matrix A is $m \times n$. Assume that (2.7) represents solution form for this LP.

Now we consider a perturbation: choose a constraint i (with $1 \leq i \leq m$), and replace $B[i]$ by $B[i] + \delta$ for form a perturbed right side B'. The form (2.6), with B' instead of B, will be called the *perturbed LP*. Use the same row operations on the perturbed LP as were used on the original LP. The result will look very much like (2.7), with perturbed right sides:

$$(2.11) \qquad \begin{pmatrix} X^T & S^T & Z & = \\ L & M & \mathbb{O}_{m \times 1} & N' \\ D & E & 1 & z_1' \end{pmatrix}$$

Proposition 2.5 relates (2.11) to the perturbed LP; we need two of the equations from that result:

$$(2.12) \qquad M \cdot B' = N', \quad E \cdot B' + z_0 = z_1'$$

The second equation gives the shadow price; the original objective value is z_1 and the perturbed value is z_1'. Thus,

$$z_1' - z_1 = E \cdot B' + z_0 - (E \cdot B + z_0) = E \cdot (B' - B)$$

Because $B' - B$ consists of 0's except for δ in the i-th position. It follows that $E \cdot (B' - B) = E[i] \cdot \delta$. This shows that $E[i]$ is the shadow price! The number $E[i]$ is the coefficient of $S[i]$ in the solution form. Also recall that $-E[i]$ is the value of the i-th variable in a solution to the dual LP.

The other equation in (2.12) gives us the allowable values: for (2.11) to be canonical, we need $N' \geq \mathbb{O}$. From (2.12), this is $M \cdot B' \geq N'$. We have

$$\mathbb{O} \leq N' = M \cdot B' = M \cdot (B' - B) + M \cdot B = M \cdot (B' - B) + N$$

The form of B' shows that $M \cdot (B' - B)$ is δ times the i-th column of M. The numbers in this column are the coefficients of $S[i]$ in the constraint part of solution form.

Summary.

PROPOSITION 2.9. *Suppose we have an LP in primal form: minimize Z such that $Z + C \cdot X = z_0$ and $A \cdot X \leq B$ and $X \geq \mathbb{O}$, where X holds the problem variables. Let S be a column of non-negative slack variables, so that the constraint is $A \cdot X + S = B$. Then when a perturbation of the ith resource bound is made, the minimum objective of the perturbed problem changes at a rate equal to the shadow price of the i-th constraint: the coefficient in the solution form of the slack variable $S[i]$. This shadow price is the negative of the value of the i-th variable of the program dual to the LP in a solution to that program. A perturbation δ is allowable if $M_i \cdot \delta + N \geq \mathbb{O}$, where M_i is the $S[i]$-constraint column in solution form and N is the right side in solution form.*

An instance worth noting: if the slack variable $S[i]$ is basic in the solution form, then $E[i] = 0$, and so changes in the resource bound $B[i]$ do not change the minimum objective.

Recall Proposition 2.6 giving us the Lagrange Multipliers for an LP. Notice that the shadow price $E[i]$ is one coordinate of the Lagrange Multipliers.

There is a technical point that needs to be mentioned: the shadow price may be ambiguous! This is because there may be choices involved in finding a solution form. Equivalently, in the dual LP, the solution may occur at more than one vector. To be more precise, we should say that the shadow price belongs to those choices, and not just to the objective.

Example. Consider this LP: maximize $x + 4y$ such that $2x + y \geq 2$ and $2x + 3y \leq 6$ and $x, y \geq 0$. We are interested in the shadow price for the first constraint. Let α be the slack variable for the first constraint, and β for the second. We will solve this LP by hand in two ways.

(a) Begin Phase I by replacing basic variable α by free variable y (in the first constraint). Phase II will follow next, and here is the solution form.

$$
\begin{pmatrix}
x & y & \alpha & \beta & = \\
2/3 & 1 & 0 & 1/3 & 2 \\
-4/3 & 0 & 1 & 1/3 & 0 \\
-5/3 & 0 & 0 & -4/3 & -8
\end{pmatrix}
$$

Read the shadow price.

(b) Begin Phase I by replacing β by y in the second constraint. Phase 1 will need one more replacement, and the resulting canonical form is in solution form:

$$
\begin{pmatrix}
x & y & \alpha & \beta & = \\
1 & 0 & -3/4 & -1/4 & 0 \\
0 & 1 & 1/2 & 1/2 & 2 \\
0 & 0 & -5/4 & -7/4 & -8
\end{pmatrix}
$$

Read the shadow price, notice that it is different than in (a)!

■

In the previous example, the dual objective equation is $Z' - 2 \cdot u + 6 \cdot v = 0$, where u, v are the dual variables. The two different solution forms give rise to two different pairs u, v: we get $u = 0, v = 4/3$ in (a) and $u = 5/4, v = 7/4$ in (b). These vectors give the same value of Z'.

◇ **Problem 36**

★ Re-solve the chair-production problem on p.29, and obtain the shadow prices for the production upper bounds. (We will discuss in class how to obtain the shadow prices in a spreadsheet solution.) Choose one of the

bounds and change it within the allowable range. Re-solve the problem, and observe that the new maximum has changed in accord with the shadow price.
◇

We have discussed perturbing the resource bounds – the constraint right sides. What about the objective coefficients? To consider perturbations of these coefficients, we can use duality to write them as resource bounds and revert to the problem already done! We will be content with this as an idea, without completing the technical details.

We have discussed shadow prices for primal LP's. Notice that equation constraints split into two inequality constraints, so an equation constraint gives rise to a pair of shadow prices. The shadow price of the equation itself can be taken to be the difference between the two shadow prices from the inequalities. This difference can be done in two ways – there is inherent ambiguity because the equation itself can be multiplied by -1 to reverse all the coefficients and the resource bound.

◇ Problem 37
Consider the problem: maximize $x + y$ such that $2x + 3y = 18$ and $7x + 6y \leq 42$ and $x, y \geq 0$. Solve the LP using the Simplex Algorithm by hand, and observe the two shadow prices on the equation.
◇

◇ Problem 38
★ We have hens, fertilized eggs, and edible eggs. The numbers of each change over time, and we want to keep track of them over 5 time intervals. (A time interval is somewhere between 10 days and 2 weeks – it doesn't matter.) During a given interval, a hen is in one of three mutually exclusive categories: laying 12 edible eggs, laying and hatching 2 fertilized eggs, or being sold. The edible eggs are sold at the end of each time interval for 1 c-unit each, and we have to have at least 70 eggs to sell at the end of each

time interval. Eggs fertilized in time interval t are mature hens[13] at the beginning of time interval $t + 2$ – this increases our stock of hens. Hens sold disappear; we get 120 c-units for each hen. At the beginning, we have 10 hens, no edible eggs, and no fertilized eggs. At the end of the 5 time intervals, we need to have at least 20 hens. How should we assign the hens to lay edible eggs, hatch fertilized eggs, or be sold during each of the 5 time intervals in order to maximize the revenue (in c-units)? Collect the shadow prices to investigate the effect of starting with more than 10 hens.
◇

7. Applications

The applications of linear programming are manifold. In addition to the application problems we have already introduced, we want to give some general types of application. In each case, we will begin with a specific problem, place it in the appropriate general context, and prove a relevant theorem.

Zero sum linear games. Each of two players writes a 1 or a 2 on a slip of paper that the other player cannot see. Next, each player writes down a guess as to the sum of the numbers written down first by both players. Each player shows the other what he/she has written down: If one player guesses correctly and the other does not, then the correct guesser receives, from the other player, a number of pennies equal to the total. In all other cases, neither player receives anything. What strategy should each player employ to keep the wins and losses even?

You might wonder what is meant by a *strategy*. If a player writes down a 1 as first number, then that player can write down 2 or 3 as the guess, since 2 and 3 are the possible totals. If a player write down 2 as first number, then the guess is 3 or 4. Thus, the possible *plays* can be enumerated: $(1, 2)$, $(1, 3)$, $(2, 3)$, $(2, 4)$. A *strategy* is a probability distribution for these four possibilities.

---

[13]There are also fertilized eggs that hatch into roosters; are are ignoring them. The "2 hens per fertilization" is an average.

Some sort of randomness is necessary, since if one player can predict with certainty what the other player will do, that first player can always win.

◇ Problem 39

Given a strategy for each player (each has four probabilities that add up to 1), what is the expected number of pennies paid out to each?

◇

The expected numbers of pennies is a sum of products of numbers of pennies and probabilities from each player's strategy. This sort of game is a *linear game*. Notice also that the expected numbers of pennies paid to one player is the negative of the expected number paid to the other – the expected payoffs add to 0. This gives rise to the term *zero sum game*. We have used a simple example; zero sum linear games show up in a variety of applied subjects and in a variety of guises.

Now we will describe the zero sum linear games abstractly and see that each one consists of a linear program and its dual. For each positive integer n, we define P_n to be the set of $v \in \mathbb{R}^n$ such that $v[i] \geq 0$ for $1 \leq i \leq n$ and such that

$$\sum_{i=1}^{n} v[i] = 1$$

We have in mind that v gives a probability distribution over a set of n mutually exclusive, exhaustive possibilities, so that $v[i]$ is the probability of the i-th possibility among the n choices. We call P_n the *probability space of dimension* n. It is not hard to see that P_n is closed and bounded. For what it's worth, the set is also convex (see Chapter 5).

A zero sum linear game starts with a given $m \times n$ matrix A, called the *payoff matrix*.[14] The idea is that there are two *players*, the *row player* and the *column player*. The row player has m choices (the number of rows of the payoff matrix) and the column player has n choices. If the row player chooses row i and the column player column j, then $A[i, j]$ gives the *payoff* to the row

---

[14]This subject is usually introduced with a 2×2 payoff matrix. Notice that the payoff matrix does not have to be square.

player. The number $-A[i,j]$ is the payoff to the column player. The fact that the row player's gain is the column player's loss is what is meant by saying that this is a *zero sum* game.[15]

We imagine that each player chooses according to a probability distribution. Given $u \in P_m$ (given a probability distribution for the m choices that the row player has) and $v \in P_n$ (a probability distribution for the n choices that the column player has), we define $f(u,v) = u^T \cdot A \cdot v$ to be the *payoff* for u, v. Notice that f is real-valued; it represents the expected payoff to the row player under the probability distributions chosen by each player.

Given a choice $v \in P_n$ by the column player, we assume that the row player tries to find $u \in P_m$ to maximize $f(u,v)$. We also assume that the column player is intelligent and chooses $v \in P_n$ to minimize this maximum. (It would be best for the column player if this minimum is negative, so that the column player gets a positive payoff.) Thus, we consider the following problem.

Min/Max problem.   *Given $m \times n$ matrix A, for each $v \in P_n$, let $\mu(v)$ be the maximum of $u^T \cdot A \cdot v$ over all $u \in P_m$. Find v to minimize $\mu(v)$.*

Given $v \in P_n$, the function $f(u,v)$ maps P_m to $\mathbb{R}$. Because P_m is closed and bounded and f is continuous, the Extreme Value Theorem[16] says that f has a maximum. Thus, $\mu(v)$ is defined for each $v \in P_n$. We will see that μ does have a minimum.

We have stated the Min/Max problem from the point of view of the column player. We could have chosen the point of view of the row player. If the row player chooses $u \in P_m$, the column player will choose $v \in P_n$ to minimize $f(u,v)$. Thus, the row player's problem is to choose u to maximize this minimum.

Max/Min problem.   *Given $m \times n$ matrix A, for each $u \in P_m$, let $\nu(u)$ be the minimum of $u^T \cdot A \cdot v$ over all $v \in P_n$. Find u to maximize $\nu(u)$.*

---

[15]Of course, we could have that $A[i,j]$ is the payoff to the *column player*; this changes maximums to minimums and vice versa in the ensuing discussion.

[16]We are jumping the gun a bit; the statement and proof of this theorem begin on p.79.

Because P_n is closed and bounded, the function $f(u, v)$ has a minimum, and so $\nu(u)$ is defined.

Here is our fundamental theorem; it says that Min/Max and Max/Min are dual linear programs, and so their optimum values are the same: mutually intelligent players will choose strategies that agree in terms of the payoff.

THEOREM 2.10. *Let A be $m \times n$. The problems Min/Max and Max/Min are dual linear programs, both of which are feasible. The minimum of $\mu(v)$ is the maximum of $\nu(u)$.*

PROOF. Let A_i be the i-th row of A. For $v \in P_n$, we claim that $\mu(v)$ is the maximum of $A_i \cdot v$ for $1 \leq i \leq m$. Indeed, let α be this maximum. For $u \in P_m$, the fact that $u[i]$ are non-negative and add up to 1 allows us to compute that

$$u^T \cdot A \cdot v = \sum_{i=1}^{m} u[i] \cdot A_i \cdot v \leq \sum_{i=1}^{m} u[i] \cdot \alpha = \alpha$$

This shows that α is an upper bound for $u^T \cdot A \cdot v$ as u ranges over P_m. There is a row k such that $A_k \cdot v = \alpha$. Let $u[k] = 1$ and $u[i] = 0$ for all other i, so that $u \in P_m$, and compute that $u^T \cdot A \cdot v = \alpha$. This proves that $\alpha = \mu(v)$.

Now can state a linear program equivalent to the Min/Max problem. In the following, v is the strategy for the column player, and the minimum y is the pay-off to the row player.

Problem L1: minimize y such that $A_i \cdot v \leq y$ for $1 \leq i \leq m$ with $v \in P_n$. (The variables are v and y, where v is a clump of n variables. The variable y is not required to be non-negative.)

Notice that if $v \in P_n$, then $v, \mu(v)$ is feasible for Problem L1. Now let v, y be a solution to Problem L1. We claim that $\mu(v) = y$ and that v is a solution to the Min/Max problem. We have $A_i \cdot v \leq y$ for each i, and so y is at least as great as the maximum of the $A_i \cdot v$. This maximum is $\mu(v)$, and so $\mu(v) \leq y$. If $\mu(v) < y$, then y could be lowered to $\mu(v)$, contradicting that v, y is a solution to the linear program. Thus, $\mu(v) = y$. As for solving the Min/Max problem, suppose that $w \in P_n$ and that $\mu(w) < \mu(v)$. We know that $\mu(w)$ is the maximum of $A_i \cdot w$ for $1 \leq i \leq m$. Then $w, \mu(w)$ is feasible for Problem L1, and this contradicts that v, y is a solution to that program. Thus, $\mu(v)$ is minimal.

We give the Max/Min problem the same treatment. Let $A^{(j)}$ be the j-th column of A for $1 \leq j \leq n$. For $u \in P_m$, we have that $\nu(u)$ is the minimum of $u^T \cdot A^{(j)}$ for $1 \leq j \leq n$.

Problem L2: maximize y' such that $u^T \cdot A^{(j)} \geq y'$ for $1 \leq j \leq n$ where $u \in P_m$.

It is easy to see that Problem L2 is feasible and that if u, y' is a solution, then $y' = \nu(u)$ solves the Max/Min problem.

We show that Problem L1 and Problem L2 are dual to each other. To see this, we put Problem L1 in primal form. The arbitrary variable y can be written $x - z$ where x, z are non-negative. The equation $\sum_{j=1}^{n} v[j] = 1$ is expressed as two inequalities. It will help to define J_k, for each positive integer k, to be the $k \times 1$ matrix all of whose entries are 1. Here is the primal form of Problem L1.

$$\text{Minimize} \quad Z \quad \text{where} \quad Z + \begin{pmatrix} \mathbb{O}_{1\times n} & -1 & 1 \end{pmatrix} \cdot \begin{pmatrix} v \\ x \\ z \end{pmatrix} = 0$$

and, using variables v, x, z, we have

$$\begin{pmatrix} A & -J_m & J_m \\ J_n^T & 0 & 0 \\ -J_n^T & 0 & 0 \end{pmatrix} \cdot \begin{pmatrix} v \\ x \\ z \end{pmatrix} \leq \begin{pmatrix} \mathbb{O}_{m\times 1} \\ 1 \\ -1 \end{pmatrix}$$

The dual of this primal form is this:

$$\text{Maximize} \quad Z' \quad \text{such that} \quad Z' + \begin{pmatrix} v' & x' & z' \end{pmatrix} \cdot \begin{pmatrix} \mathbb{O}_{m\times 1} \\ 1 \\ -1 \end{pmatrix} = 0$$

and

$$\begin{pmatrix} v' & x' & z' \end{pmatrix} \cdot \begin{pmatrix} A & -J_m & J_m \\ J_n^T & 0 & 0 \\ -J_n^T & 0 & 0 \end{pmatrix} \geq \begin{pmatrix} \mathbb{O}_{1\times n} & -1 & 1 \end{pmatrix}$$

(The variables are v', x', z' and v' is $1 \times m$.)

We claim that this dual problem is Problem L2. Indeed, the inequality constraints on v' show that $v'^T \in P_m$. Also, $v' \cdot A^{(j)} + x' - z' \geq 0$ for $1 \leq j \leq n$ shows that $z' - x'$ is less than or equal to $\nu(v')$. Maximizing $z' - x'$ would give the maximum of $\nu(v')$.

We have proved that Min/Max and Max/Min are dual to each other. We also know that they are each feasible. The Duality Theorem finds solutions to each and shows that the minimum of the primal problem (Min/Max) is the maximum of the dual problem (Max/Min). ∎

◇ **Problem 40**

★ Solve the problem stated at the beginning of this section.

◇

Fractional Objectives. A baseball player describes the various ways he can approach batting as an element S of $\mathbb{R}^5$. Each coordinate represents a factor that he can control and that takes on a numerical value between 0 and 1 inclusive. He estimates that the vector S will result in $375 + \alpha \cdot S$ at-bats[17] and $105 + \beta \cdot S$ hits over a hundred game period, where the vectors α, β are given in the table below. Find S to maximize his batting average (ratio of hits to at-bats).

α	-20	8	10	-7	9
β	8	-4	7	10	-11

Here is a general model of a problem like the batting problem. We are given positive integers m, n, a $1 \times n$ matrix C, a real number c_0, a $1 \times n$ matrix E, a real number e_0, an $m \times n$ matrix A, an $m \times 1$ matrix B. We assume that if $A \cdot X \leq B$, then $E \cdot X + e_0 > 0$. Define

$$f(X) = \frac{C \cdot X + c_0}{E \cdot X + e_0}$$

Fractional Problem: minimize $f(X)$ such that $A \cdot X \leq B$ and $X \geq \mathbb{O}$.

---

[17]Regarding the variation in at-bats: given roughly 400 plate appearances, what S actually controls is the number of walks the player will likely get.

Note: The assumption that $E \cdot X + e_0$ is positive prevents $f(X)$ from having 0 in the denominator.

PROPOSITION 2.11. *With the notation just given, consider the following LP: minimize $C \cdot Y + c_0 \cdot Y_0$ such that $A \cdot Y - B \cdot Y_0 \leq \mathbb{O}$ and $E \cdot Y + e_0 \cdot Y_0 = 1$ and $Y \geq \mathbb{O}$ and $Y_0 \geq 0$. The Fractional Problem has a solution if and only if the LP has a solution in which $Y_0 > 0$. If the LP has solution $Y = V$ and $Y_0 = V_0$, then $X = V/V_0$ is a solution to the Fractional Problem.*

PROOF. We will write FP for the Fractional Problem. The objective for the FP is $f(X)$. We will write the LP problem variables (Y, Y_0) and its objective as $Z(Y, Y_0)$.

We show how to go back and forth between the two problems.

Claim 1. Let V be feasible for the FP. Write $V_0 = E \cdot V + e_0$, and then $(V/V_0, 1/V_0)$ is feasible for the LP, and $f(V) = Z(V/V_0, 1/V_0)$.

Proof. Since V is non-negative and V_0 positive, the vectors V/V_0 and $1/V_0$ are non-negative. As to the equations: we have $A \cdot V \leq B$, and so $A \cdot V/V_0 - B/V_0 \leq \mathbb{O}$, and we have $E \cdot V/V_0 + e_0/V_0 = (E \cdot V + e_0)/V_0 = 1$ by the definition of V_0. Thus, $(V/V_0, 1/V_0)$ is feasible for the LP.

Claim 2. Let (U, U_0) be feasible for the LP, and suppose that $U_0 > 0$. Then U/U_0 is feasible for the FP, and $Z(U, U_0) = f(U/U_0)$.

Proof. We have $A \cdot U \leq B \cdot U_0$, and so since $U_0 > 0$, we have $A \cdot U/U_0 \leq B$.

Claim 3. Let $(U, 0)$ be feasible for the LP, and let V be feasible for the FP. Define V_0 as in Claim 1, so that $(V/V_0, 1/V_0)$ is feasible for the LP. Then $V + U$ is feasible for the FP. If V_0 is defined as in Claim 1 and if

$$f(V) \leq f(V + U)$$

then $Z(V/V_0, 1/V_0) \leq Z(U, 0)$.

Proof. Since $V \geq \mathbb{O}$ and $U \geq \mathbb{O}$, we have $V + U \geq \mathbb{O}$. We have

$$A \cdot U - B \cdot 0 \leq \mathbb{O} \quad \text{and} \quad A \cdot V \leq B$$

and so

$$A \cdot (V + U) = A \cdot V + A \cdot U \leq B + \mathbb{O} = B$$

This shows that $V + U$ is feasible for the FP. Define V_0 as in Claim 1, and assume that $f(V) \le f(V + U)$. This is

$$\frac{C \cdot V + c_0}{E \cdot V + e_0} \le \frac{C \cdot (V + U) + c_0}{E \cdot (V + U) + e_0}$$

Since $E \cdot U = 1$ and $E \cdot V + e_0 > 0$, this is

$$\left(C \cdot V + c_0\right) \cdot \left(E \cdot V + e_0 + 1\right) \le \left(C \cdot V + c_0 + C \cdot U\right) \cdot \left(E \cdot V + e_0\right)$$

Canceling $(C \cdot V + c_0) \cdot (E \cdot V + e_0)$ from both sides, we get

$$C \cdot V + c_0 \le C \cdot U \cdot (E \cdot V + e_0)$$

Dividing by the positive number $E \cdot V + e_0$, we see that $f(V) \le Z(U)$. Since $f(V) = Z(V/V_0, 1/V_0)$, Claim 3 holds.

Now we can begin the proof proper. Suppose that V is a solution to the FP. Define V_0 as in Claim 1, so that $(V/V_0, 1/V_0)$ is feasible for the LP, and we have $Z(V/V_0, 1/V_0) = f(V)$. We claim that $(V/V_0, 1/V_0)$ is a solution to the LP. Indeed, let (U, U_0) be feasible for the LP. If $U_0 = 0$, then Claim 3 shows that $V + U$ is feasible for the FP. Since $f(V)$ is the minimum, we have $f(V) \le f(V + U)$, and Claim 3 implies that $Z(V/V_0, 1/V_0) \le Z(U, 0)$. If $U_0 > 0$, then Claim 2 says that U/U_0 is feasible for the FP, and $Z(U, U_0) = f(U/U_0)$. Then

$$Z(U, U_0) = f(U/U_0) \ge f(V) = Z(V/V_0, 1/V_0)$$

We have proved that the LP has a solution with $Y_0 > 0$.

For the converse, suppose that the LP has a solution (U, U_0) with $U_0 > 0$. Claim 2 says that U/U_0 is feasible for the FP, and $f(U/U_0) = Z(U, U_0)$. We claim that U/U_0 solves the FP. If V is feasible for the FP, then Claim 1 shows that $(V/V_0, 1/V_0)$ is feasible for the LP, and $Z(V/V_0, 1/V_0) = f(V)$. Compute

$$f(V) = Z(V/V_0, 1/V_0) \ge Z(U, U_0) = f(U/U_0)$$

as needed.   ■

◇ **Problem 41**
★ Solve the batting problem.

Allocation Over a Graph. Warehouse A has 9 tables, and warehouse B has 8. Store α needs 11 tables shipped to it, store β needs 6. We imagine all the ways that the tables can be shipped from the two warehouses to the two stores. Let $A\alpha$ be the number of tables shipped from A to α, and, similarly, create variables $A\beta$, $B\alpha$, $B\beta$.

 Problem 42

What are the equations on these four variables?

The equations just derived allow real number values in the variables; we almost certainly want to ship *integer* numbers of tables.

 Problem 43

Show that the basic vector solutions to the constraint equations all have integer values.

This last result is suspicious. Linear equations with integer coefficients may have basic vectors with non-integer entries. We will show that the warehouse constraints have a special form: the coefficient matrix is the *incidence matrix* of a graph. Here are the relevant definitions. A *directed graph* G consists of a finite set V of *vertices*, and a finite set E of *edges*. We think of the vertices as *points* and we imagine that each of the edges "connects" two vertices – a *tail* to a *head.*. This pictorial view is quite helpful, but we need to turn to formal terms: A graph G consists of a pair of finite sets (V, E) and two functions $t : E \to V$ and $h : E \to V$, such that $t(e) \neq h(e)$ for all $e \in E$. For each edge e, the vertex $t(e)$ is the *tail* of the edge, and $h(e)$ is the *head* of the edge.

Suppose we have a graph with m vertices and n edges. An *incidence matrix* of the graph is an $m \times n$ matrix A that we will now describe. Each row of A corresponds to a vertex, and each column to an edge. For edge e we have $A[t(e), e] = -1$ and $A[h(e), e] = 1$. All other entries in column e of A are 0. Notice that the 1 indicates the vertex at the head of e, and the -1 the vertex at the tail. We think of the tail to the head as the *direction* of the edge e.

In manipulating incidence matrices, it will be useful to allow there to be extra columns that are all 0. Thus, there may be columns of an incidence matrix that do not correspond to any of the edges.

Here is our main theorem; its novelty consists in the fact that it guarantees *integer* solutions to a system of equations.

PROPOSITION 2.12. *Let A be the incidence matrix of a graph, and suppose that A is $m \times n$. Let B be an $m \times 1$ matrix having integer entries. Then every basic vector for $A \cdot X = B$ has integer entries.*

PROOF. As above, let V be the set of vertices and E the set of edges. For each edge e we have an unknown $X[e]$ in the system of equations.

For some basic vector, choose an edge $f \in E$ such that $X[f]$ is a basic variable. Then $A[h(f), f)] = 1$. We will use $X[f]$ as a basic variable; to do this, we need to eliminate the occurrence of $X[f]$ in all other equations (rows of A). The only other row having $X[f]$ is row $t(f)$, and $A[t(f), f] = -1$. Thus, to eliminate $X[f]$ from the other equations, we need to add row $h(f)$ to row $t(f)$. When this is done, the other equations have a coefficient matrix A' that is the incidence matrix of another graph. Let X' be X with $X[f]$ deleted, and then the equation $A \cdot X = B$ becomes[18]

$$X[f] = B[h(f)] - \sum_{e \neq f} A[h(f), e] \cdot X[e] \quad \text{and} \quad A' \cdot X' = B'$$

By induction, every basic vector solution for $A' \cdot X' = B'$ has integer values. The equation for $X[f]$ shows that this variable will have an integer value as well. ■

Suppose we are given an LP: minimize $Z + C \cdot X = z_0$ such that $A \cdot X = B$ and $X \geq \mathbb{O}$. Furthermore, suppose we require that the entries in X be integers. To solve this problem, we might ignore the constraint that X be integers, and apply the simplex algorithm as is. If the solution comes out to have integer entries, then we obtain the integer constraint for free. But what if the simplex solution comes out to have some non-integer entries? Can we

---

[18]This process is most easily seen in specific examples.

just round to the nearest integers to obtain a solution to the integer problem? The answer is, "No!" Integer problems are, in general, substantially more complicated than their real number cousins.[19] We will not pursue this except to mention that Proposition 2.12 gives a setting in which the simplex algorithm will automatically find a solution with integer entries – when the coefficient matrix is the incidence matrix of a graph. This situation comes up in several classes of applied problems.

◇ **Problem 44**
★ In the warehouse problem, suppose that the table to the left gives the shipping costs per table from each warehouse to each store. Determine how to ship the tables to minimize the total shipping cost.
◇

Costs

	A	B
α	3	4
β	2	5

Effectiveness

	A	B	C	D
1	2	2	4	0
2	1	3	5	0
3	0	0	1	6
4	5	0	2	6

◇ **Problem 45**
★ Each of four workers A,B,C,D needs to be assigned to one of four tasks 1,2,3,4. The table to the right abovbe gives a measure of effectiveness of each worker for each task. We will show how to assign workers to tasks to maximize the total effectiveness. For each worker i and task j, have a variable $X[i \to j]$. Effectiveness is the sum of the variables multiplied by coefficients from the table. The constraints are these: for each worker i, the sum of $X[i \to j]$ over all j is 1; and for each task j, the sum of $X[i \to j]$ over all i is 1. Solve this problem.
◇

---

[19]For instance, the *travelling salesman problem* of finding the shortest path through a graph visiting every vertex seems to require substantially more computing time than an LP with the same number of variables.

CHAPTER 3

Multivariate Calculus

1. Open Sets and Continuous Functions

To state and prove each of our main theorems on nonlinear optimization, we need some apparatus from topology and analysis. As with the linear algebra that supported our work on LP's, we will move with some speed, trying to keep close to the facts we will actually use. We will occasionally linger over ideas that are basic to advanced mathematics generally.

The elements of $\mathbb{R}^n$ will be called points or vectors, interchangeably. We have already adopted the idea that elements of $\mathbb{R}^n$ are $n \times 1$ matrices. And we know how to calculate dot products and norms. Recall, for $v \in \mathbb{R}^n$, that $|v|^2 = v \circ v = v^T \cdot v$.

The *Triangle Inequality* is basic to estimation.

TRIANGLE INEQUALITY. *Let* $x, y \in \mathbb{R}^n$. *Then* $|x + y| \le |x| + |y|$.

PROOF. Using the Cauchy-Schwarz inequality (Proposition 1.8), we calculate

$$
\begin{aligned}
|x + y|^2 &= (x + y) \circ (x + y) \\
&= x \circ x + 2 \cdot x \circ y + y \circ y \qquad \text{linearity of dot product} \\
&= |x|^2 + 2 \cdot x \circ y + |y|^2 \\
&\le |x|^2 + 2 \cdot |x| \cdot |y| + |y|^2 \qquad \text{by Cauchy-Schwarz} \\
&= \left(|x| + |y| \right)^2
\end{aligned}
$$

We end up with $|x + y|^2 \le (|x| + |y|)^2$. Taking square roots, we obtain the Triangle Inequality. ∎

Topology is the study of certain naturally occurring distinguished subsets of a given set that are said to be *open* in the larger set. The fundamental open subset of $\mathbb{R}^n$ is the *open disk*: for $x \in \mathbb{R}^n$ and a positive number r, the *open disk about x of radius r* is the set of $y \in \mathbb{R}^n$ such that $|y - x| < r$. We denote this set $B(x, r)$. When $n = 1$, the set $B(x, r)$ is the open interval $(x - r, x + r)$ on the real line. When $n = 2$, the set $B(x, r)$ is the interior of a circle centered at x of radius r. When $n = 3$, the set $B(x, r)$ is the interior of a sphere.

The open sets in $\mathbb{R}^n$ are unions of disks: a subset V of $\mathbb{R}^n$ is *open* if for every $x \in V$, there is a positive real number r such that $B(x, r) \subseteq V$.

Here are the principal properties of open sets.

PROPOSITION 3.1. *We have the following.*

(a) If $x \in \mathbb{R}^n$ and $r > 0$, then $B(x, r)$ is open.
(b) The set $\mathbb{R}^n$ is an open subset of itself.
(c) The empty set is an open subset of $\mathbb{R}^n$.
(d) An arbitrary union[1] of open subsets of $\mathbb{R}^n$ is an open subset of $\mathbb{R}^n$.
(e) If U, V are open subsets of $\mathbb{R}^n$, then $U \cap V$ is open.

◇ Problem 46

Let A be a $1 \times n$ matrix, and let $b \in \mathbb{R}$. Let V be the set of $X \in \mathbb{R}^n$ such that $A \cdot X < b$. Prove that V is open. (Hint: If $A = \mathbb{O}$, then V is either $\mathbb{R}^n$ or empty; if $A \neq \mathbb{O}$, let $X \in V$, define $r = (b - AX)/|A|$ and show that $B(X, r) \subseteq V$.)
◇

◇ Problem 47

Let A be an $m \times n$ matrix, and let B be $m \times 1$. Prove that the set of $X \in \mathbb{R}^n$ such that $A \cdot X < B$ is open. (Hint: this set is a finite intersection of sets of the type of the previous problem.)

---

[1]Arbitrary union: we have a set I (can be literally anything), and for each $\alpha \in I$, we have an open set V_α in $\mathbb{R}^n$. The union is the set of $x \in \mathbb{R}^n$ such that there is $\alpha \in I$ with $x \in V_\alpha$.

◇ **Problem 48**

Let n be a positive integer, and let G be the set of invertible $n \times n$ matrices. Then G is open in the set of all $n \times n$ matrices. (Hint: Proposition 1.12.)
◇

We turn to the study of functions. If $V \subseteq \mathbb{R}^n$ and $p \in V$, then the function $f : V \to \mathbb{R}^m$ is *continuous at* p if, for all $\epsilon > 0$, there is $\delta > 0$ such that if $w \in V$ with $|w - p| < \delta$, then $|f(w) - f(p)| < \epsilon$. It is no coincidence that this definition is, word for word, the definition in the case of a function of one variable, with the norm on $\mathbb{R}^n$ playing the role of absolute value (norm!) on the reals.[2] You might recall that the δ, ϵ definition of continuity is equivalent to the limit definition usually given in elementary calculus.

If $f : V \to \mathbb{R}^m$, we say that f is *continuous on* V if it is continuous at every $p \in V$.

◇ **Problem 49**

Define $h : \mathbb{R}^n \to \mathbb{R}$ by $h(x) = |x|$. Show that h is continuous.
◇

◇ **Problem 50**

Let $p \in \mathbb{R}^n$ and $r > 0$. Let $f : B(p, r) \to \mathbb{R}^m$. Suppose that there is $\delta > 0$ such that if $x \in B(p, r)$, then $|f(x) - f(p)| \leq \delta \cdot |x - p|$. Show that f is continuous at p. (Note: the condition here is identified by saying that f has a Δ-bound at p.)
◇

◇ **Problem 51**

Let A be an $m \times n$ matrix, and define $f : \mathbb{R}^n \to \mathbb{R}^m$ by $f(X) = A \cdot X$. Show that $f(X)$ is continuous. (Hint: Lemma 1.9.)
◇

---

[2]See [**4**] or any other standard text introducing advanced calculus.

A function $f : V \to \mathbb{R}^m$ can be thought of as a bundle of functions, one for each coordinate. For each $x \in V$, write $f_j(x)$ for the j-th coordinate of $f(x)$. We have

$$f(x_1, x_2, \ldots, x_n) = \begin{pmatrix} f_1(x_1, x_2, \ldots, x_n) \\ f_2(x_1, x_2, \ldots, x_n) \\ \vdots \\ f_m(x_1, x_2, \ldots, x_n) \end{pmatrix}$$

We call f_f the i-th *coordinate function* of f.

PROPOSITION 3.2. *Let V be an open subset of $\mathbb{R}^n$ and let $f : V \to \mathbb{R}^m$, with coordinate functions f_j, for $1 \leq j \leq m$. Then f is continuous if and only if each f_j is continuous.*

PROOF. Suppose that each f_j is continuous. Let $p \in V$, and we will show that f is continuous at p. Let $\epsilon > 0$. For each j, there is a positive real number δ_j such that if $y \in V$ and $|y - x| < \delta_j$, then

$$|f_j(y) - f_j(x)| < \epsilon$$

Let δ be the minimum of the δ_j, for $1 \leq j \leq m$, so that δ is a positive real number. If $y \in V$ and $|y - x| < \delta$, then $|y - x| < \delta_j$ for all j, and we compute

$$|f(y) - f(x)|^2 = \sum_{j=1}^{m} |f_j(y) - f_j(x)|^2 \leq m \cdot \epsilon^2$$

This implies that f is continuous at p.

We leave the converse to you.   ■

◇ Problem 52

Let V be an open subset of $\mathbb{R}^n$ and let $f : V \to \mathbb{R}^m$ be continuous. Then each coordinate function f_j is continuous.

◇

Usually, the nature of the formula for a function makes it obvious that the function is continuous. In other words, we don't do specific examples from the definition, as a rule. The proofs of the following statements are similar to those given in standard analysis texts for the case of one variable. We might have time to give sample arguments in class.

PROPOSITION 3.3. *Let $V \subseteq \mathbb{R}^n$ and suppose that $f : V \to \mathbb{R}^m$ and $g : V \to \mathbb{R}^m$ are both continuous. Then*

(a) If $a \in \mathbb{R}$, then $a \cdot f(x)$ is continuous.
(b) $f(x) + g(x)$ is continuous.
(c) $f(x) \circ g(x)$ is continuous. (dot product)
(d) If $m = 1$ and $g(x) \neq 0$ for all $x \in V$, then $1/g(x)$ is continuous.

Since $f(x)/g(x) = f(x) \cdot (1/g(x))$, it follows that the ratio of continuous functions is continuous, wherever the denominator is not 0. Thus, Proposition 3.3 shows that any rational combination of variables is continuous. For instance, all polynomial functions, rational functions, and algebraic functions are continuous wherever they are defined.

Inequalities with continuous functions give open sets.

PROPOSITION 3.4. *Let $f : \mathbb{R}^n \to \mathbb{R}^m$ be continuous and let $b \in \mathbb{R}^m$. Define V to be the set of $x \in \mathbb{R}^n$ such that $f(x) < b$. Then V is an open subset of $\mathbb{R}^n$.*

PROOF. Let $f_j(x)$, for $1 \leq j \leq m$, be the coordinate functions of f. Define V_j to be the set of $x \in \mathbb{R}^n$ such that $f_j(x) < b[j]$. Then

$$V = \bigcap_{j=1}^{m} V_j$$

We will show that V is open by showing that each of the V_j is open. By Proposition 3.2, each of the coordinate functions is continuous. We have reduced the proof to the case $m = 1$!

Assume that $m = 1$, so that $f(x)$ is a real number. Let $x \in V$ and we must find $r > 0$ such that $B(x, r) \subseteq V$. We will see that the r is a "δ" as in the definition of continuity. For "ϵ" take $b - f(x)$; because $x \in V$, this is a positive number.

Since f is continuous at x, there is $\delta > 0$ such that if $y \in \mathbb{R}^n$ and $|y-x| < \delta$, then we have

$$|f(y) - f(x)| < b - f(x)$$

Then

$$f(y) = f(y) - f(x) + f(x) \leq |f(y) - f(x)| + f(x) < b - f(x) + f(x) = b$$

This proves that $f(y) < b$, so that $y \in V$. We have proved that $B(x, \delta) \subseteq V$, and so V is open. ∎

You probably remember that a composite of continuous function is continuous. As with Proposition 3.3, the proof is identical with that given in the one variable case. But this argument is so basic that we can't help giving it here.

PROPOSITION 3.5. *Let* $V \subseteq \mathbb{R}^n$ *and* $W \subseteq \mathbb{R}^m$ *and let* $f : V \to W$ *be continuous, and let* $g : W \to \mathbb{R}^k$ *be continuous. Then* $g(f(x))$ *is continuous.*

PROOF. Notice that the domain of $g(f(x))$ is V. Let $v \in V$ and $\epsilon > 0$. Then $f(v) \in W$, and since g is continuous on W, there is $\delta > 0$ such that if $y \in W$ and $|y - f(v)| < \delta$, then $|g(y) - g(f(v))| < \epsilon$.

Since f is continuous at v, we can use the number δ as our "ϵ" and find $\mu > 0$ such that if $x \in V$ and $|x - v| < \mu$, then $|f(x) - f(v)| < \delta$.

Now we claim that if $|x - v| < \mu$ with $x \in V$, then $|g(f(x)) - g(f(v))| < \epsilon$. Indeed, the previous paragraph shows that $|f(x) - f(v)| < \delta$. We know that $f(x) \in W$, and so the definition of δ shows that $|g(f(x)) - g(f(v))| < \epsilon$, as needed. ∎

You know from calculus that the root functions, such as $\sqrt{x}$, the trigonometric functions, the exponential functions, the logarithmic functions, and functions defined by power series are continuous wherever they are defined. Proposition 3.3 and Proposition 3.5 then say that any combination, algebraic or composite, of such functions is continuous.

PROPOSITION 3.6. *Let n be a positive integer, and let G be the set of invertible $n \times n$ matrices. Define $f : G \to G$ by $f(A) = A^{-1}$. Then f is continuous.*

PROOF. Let $A, B \in G$, and use Proposition 1.10 shows that[3]

$$|A \cdot (A^{-1} - B^{-1}) \cdot B| \geq \mu(A) \cdot |A^{-1} - B^{-1}| \cdot \mu(B^T)$$

Proposition 1.11 shows that $\mu(A)$ and $\mu(B^T)$ are not 0. On the other hand,

$$A \cdot (A^{-1} - B^{-1}) \cdot B = B - A$$

and we see that

$$\frac{1}{\mu(A) \cdot \mu(B^T)} \cdot |B - A| \geq |A^{-1} - B^{-1}|$$

It follows that f is continuous. ∎

There is a very important existence theorem for maximums and minimums for functions defined on what is called the *closed* sets. A subset C of $\mathbb{R}^n$ is *closed* if it is the complement of an open set: if the set of $x \in \mathbb{R}^n$ such that $x \notin C$ is open. Using the definition of open to state this: if $x \in \mathbb{R}^n$ and $x \notin C$, then there is a positive number r such that $B(x, r)$ is disjoint from C. This implication is often used in its contrapositive form: the set C is closed if and only if it satisfies this implication: if $B(x, r) \cap C$ is non-empty for all positive numbers r, then $x \in C$.

The following facts mirror the properties of open sets stated in Proposition 3.1.

PROPOSITION 3.7. *We have the following.*

(a) Let $x \in \mathbb{R}^n$ and $r > 0$. Then $\mathbb{R}^n - B(x, r)$ is closed.
(b) The empty set is a closed subset of $\mathbb{R}^n$.
(c) The set $\mathbb{R}^n$ is a closed subset of itself.
(d) An arbitrary intersection of closed subsets of $\mathbb{R}^n$ is closed.
(e) If C, D are closed subsets of $\mathbb{R}^n$, then $C \cup D$ is closed.

---

[3]The transpose appears on B because we are pulling it off on the right.

Let $x \in \mathbb{R}^n$ and $r > 0$. Let $D(x, r)$ be the set of $y \in \mathbb{R}^n$ such that $|y-x| \leq r$. This set is called the *closed disk*.

◇ Problem 53

Let $x \in \mathbb{R}^n$ and $r > 0$. Show that $D(x, r)$ is a closed subset of $\mathbb{R}^n$.
◇

◇ Problem 54

Let $S \subseteq \mathbb{R}^n$. The *boundary* of S is the set of all $x \in \mathbb{R}^n$ such that for all $r > 0$, the set $B(x, r) \cap S$ is not empty, and the set $B(x, r) \setminus S$ is not empty. Show that the set S is closed if and only if it contains its boundary.
◇

As *strict inequalities* with continuous functions define open sets, so *equations* with continuous functions define closed sets.

◇ Problem 55

Let $f : \mathbb{R}^n \to \mathbb{R}^m$ be continuous. Let $b \in \mathbb{R}^m$. Then the set C of $x \in \mathbb{R}^n$ such that $f(x) = b$ is closed.
◇

It follows that if A is an $m \times n$ matrix and B an $m \times 1$ matrix, then the set of $X \in \mathbb{R}^n$ such that $AX = B$ is a closed set. It is also true that the set of X such that $AX \leq B$ is closed. Thus, the set of feasible vectors for an LP defines a closed set.

Open and closed sets arise very naturally in applications. We caution you that if we are given an *arbitrary* subset of $\mathbb{R}^n$, chances are it is neither open nor closed.

2. Limit Points and Extreme Values

A subset C of $\mathbb{R}^n$ is *bounded* if it is contained in some closed disk. The main theorem of this section is that a real-valued continuous function on a closed and bounded set has a minimum and a maximum. Our proof will be built up starting with the *Bolzano-Weierstrass Theorem* of one-variable analysis. Before going on, you should make sure you remember the Completeness Property (or Axiom) of the real numbers. We have mentioned [4] as a reference. The Bolzano-Weierstrass Theorem can be stated in several different ways – here's the statement we will use.

BOLZANO-WEIERSTRASS THEOREM. *Each bounded sequence of real numbers has a converging subsequence.*

PROOF. Let $a_1, a_2, \ldots$ be in the closed interval $[b, c]$, with $b \leq c$. Define $b_0 = b$ and $c_0 = c$. Let $d = (b_0 + c_0)/2$. Either there are infinitely many j such that $a_j \in [b_0, d]$, or there are infinitely many j such that $a_j \in [d, c_0]$. (Both statements can be true.) In the former case, let $b_1 = b_0$ and $c_1 = d$; in the latter case, let $b_1 = d$ and $c_1 = c_0$. Thus, there are infinitely many j such that $a_j \in [b_1, c_1]$. Choose one of these j, call it $k(1)$. There are infinitely many $j > k(1)$ such that $a_j \in [b_1, c_1]$.

Assume we have

$$b_0 \leq b_1 \leq \cdots \leq b_n < c_n \leq \cdots \leq c_1 \leq c_0$$

with $k(1) < \cdots < k(n)$ and $a_{k(i)} \in [b_i, c_i]$ for each i. Assume also that there are infinitely many $j > k(n)$ such that $a_j \subset [b_n, c_n]$. Let $d = (b_n + c_n)/2$, and use either $[b_n, d]$ or $[d, c_n]$ as $[b_{n+1}, c_{n+1}]$, such that there are infinitely many j with $a_j \in [b_{n+1}, c_{n+1}]$. Choose $k(n+1) > k(n)$ with $a_{k(n+1)} \in [b_{n+1}, c_{n+1}]$. This gives us a sequence $a_{k(j)}$ with properties as just enunciated. We see that $0 \leq c_n - b_n = (c_0 - b_0)/2^n$.

The sequence b_j is increasing and bounded above. By the Monotone Convergence Theorem,[4] the sequence b_j converges, say to p. We see that $p \leq c_n$ for all n, as well, so that $p \in [b_n, c_n]$ for all n.

We claim that the subsequence $a_{k(j)}$, for $j = 1, 2, \ldots$ converges to p. Indeed, let $\epsilon > 0$; there is a positive integer n with $c_n - b_n < \epsilon$. If $j \geq n$, then $a_{k(j)} \in [b_n, c_n]$, and so $|a_{k(j)} - p| \leq c_n - b_n < \epsilon$. ∎

---

[4]This is our use of the Completeness Property of the real numbers.

The Bolzano-Weierstrass Theorem holds in $\mathbb{R}^n$ as well.

PROPOSITION 3.8. *Let n be a positive integer. Each bounded sequence in $\mathbb{R}^n$ has a converging subsequence.*

PROOF. Let $a_1, a_2, \ldots$ be a sequence in the bounded set $C \subset \mathbb{R}^n$. The boundedness of C shows that there are real numbers $b < c$ such that $a_i[j] \in [b, c]$ for all $i \geq 1$ and $1 \leq j \leq n$.

We prove the proposition by induction on n, the Bolzano-Weierstrass Theorem being the case $n = 1$. Given $n > 1$, write each $a_j = (d_j, e_j)$ where $d_j \in \mathbb{R}^{n-1}$ and $e_j \in \mathbb{R}$, for each $j \geq 1$. By induction, the sequence d_j has a converging subsequence $d_{k(j)}$, where $k(1) < k(2) < \ldots$. Say that $d_{k(j)}$ converges to $p \in \mathbb{R}^{n-1}$. By the Bolzano-Weierstrass Theorem, the sequence $e_{k(j)}$ has a converging subsequence $e_{k(h(j))}$; say the limit is $q \in \mathbb{R}$. Then the subsequence $a_{k(h(j))}$ converges to $(p, q) \in \mathbb{R}^n$, for if $\epsilon > 0$, then there is N such that if $j > N$, then $|d_{k(j)} - p| < \epsilon$, and there is M such that if $j > M$, then $|e_{k(h(j))} - q| < \epsilon$. We can make M larger, if necessary, so that $j > M$ implies that $h(j) > N$. For $j > M$, we have

$$|a_{k(h(j))} - (p, q)|^2 = |d_{k(h(j))} - p|^2 + |e_{k(h(j))} - q|^2 < \epsilon^2 + \epsilon^2$$

This does it.   ∎

Now we can take a major step toward our main theorem. The following result is important in its own right.

PROPOSITION 3.9. *Let C be a closed and bounded subset of $\mathbb{R}^n$, and suppose that $f : C \to \mathbb{R}^m$ is continuous. Then the image $f(C)$ of f is closed and bounded.*

PROOF. Assume that $f(C)$ is not bounded, and we will derive a contradiction. For each positive integer k, there is $a_k \in C$ such that $|f(a_k)| \geq k$. By Proposition 3.8, there is a subsequence $h(1) < h(2) < \ldots$ such that $a_{h(j)}$ converges to some $p \in \mathbb{R}^n$.

We claim that $p \in C$. If $p \notin C$, then since C is closed there is $r > 0$ with $B(p,r) \cap C$ empty. But the sequence $a_{h(k)}$ is eventually in $B(p,r)$, and $a_{h(k)} \in C$, a contradiction. Since $p \in C$, we can refer to $f(p)$.

The function f is continuous at p, and so there is $\delta > 0$ such that if $x \in C$ and $|x - p| < \delta$, then $|f(x) - f(p)| < 1$. For these x, we have

$$(3.1) \quad |f(x)| = |f(x) - f(p) + f(p)| \le |f(x) - f(p)| + |f(p)| < 1 + |f(p)|$$

Since $a_{h(k)} \to p$, we see that $|f(a_{h(k)})| < 1 + |f(p)|$ eventually. But $|f(a_{h(k)})| \ge h(k)$, and $h(k)$ goes to infinity. This is a contradiction; we have proved that $f(C)$ is bounded.

To see that $f(C)$ is closed, let $w \in \mathbb{R}^m \setminus f(C)$ and consider the function $g(x) = 1/|w - f(x)|$, mapping C into $\mathbb{R}$. By the continuity algebra, and because the norm is continuous, the function g is continuous on C. By what we just proved, the image $g(C)$ is bounded. Let $M > 0$ be such that $1/|w - f(x)| < M$ for all $x \in C$. Then $|w - f(x)| > 1/M$, and it follows that $f(x)$ cannot be in the open disk $B(w, 1/M)$.

We have proved this: if $w \in \mathbb{R}^m \setminus f(C)$, there is an open disk about w such that no element of $f(C)$ is in that disk. This proves that $f(C)$ is closed. ∎

EXTREME VALUE THEOREM. *Suppose that C is a closed and bounded subset of $\mathbb{R}^n$, and let $f : C \to \mathbb{R}$ be continuous. Then there are $\alpha, \beta \in C$ such that $f(\alpha) \le f(x) \le f(\beta)$ for all $x \in C$.*

PROOF. By Proposition 3.9, the set $f(C)$ is closed and bounded. By the Completeness property of the real numbers, this set has a sup s. If $s \notin f(C)$, then since $f(C)$ is closed, there is $\epsilon > 0$ such that the open interval $(s-\epsilon, s+\epsilon)$ does not intersect $f(C)$. By definition, s is the *least* upper bound of $f(C)$. Therefore, the number $s - \epsilon$ is not an upper bound, and there is $c \in C$ such that $s - \epsilon < f(c)$. Since s is an upper bound, we have $f(c) \le s$, and now $f(c)$ is an element of $f(C)$ in $(s - \epsilon, s + \epsilon)$, a contradiction. This proves that $s = f(\beta)$ for some $\beta \in C$.

Similarly, there is $\alpha \in C$ such that $f(\alpha)$ is the inf (minimum) of $f(C)$. ∎

Boundedness is crucial.

◇ **Problem 56**

Show that the function x (real variable) has no minimum on the closed set of all $(x, y) \in \mathbb{R}^2$ such that $x \cdot y \geq 1$ and $x \geq 0$.

◇

◇ **Problem 57**

Let A be an $m \times n$ matrix. Recall the sphere S_n and the functions μ and ν from p.20. Prove that there are $u, v \in S_n$ such that $|A \cdot u| = \mu(A)$ and $|A \cdot v| = \nu(A)$.

◇

3. Derivatives

Remember partial derivatives from Calculus III? Given a function f of several variables, one of which is x, the *partial derivative* $\frac{\partial f}{\partial x}$ is defined as the derivative of f with respect to x, holding all the other variables fixed.[5] Here is a more formal version. Suppose that $V \subseteq \mathbb{R}^n$ is open and $f : V \to \mathbb{R}$ is defined. For a variable x_j of f, let $e_j \in \mathbb{R}^n$ with $e_j[j] = 1$ and $e_j[i] = 0$ for all $i \neq j$. In other words, e_j is the j-th column of the identity matrix I_n. If $p \in V$, then we have the following limit definition of the partial derivative of f with respect to x_j at p.

$$(3.2) \qquad \frac{\partial f}{\partial x_j}(p) = \lim_{t \to 0} \frac{f(p + t \cdot e_j) - f(p)}{t}$$

The difference $f(p + t \cdot e_j) - f(p)$ involves a change in only the j-th coordinate.

Our typical setting will involve a function mapping into $\mathbb{R}^m$ where m can be greater than 1. Let $V \subseteq \mathbb{R}^n$ be open, and suppose we have $f : V \to \mathbb{R}^m$.

---

[5]You might give yourself a quick review from a calculus text.

For each $x \in V$, the vector $f(x)$ has m coordinates. We will write

$$f(x) = \begin{pmatrix} f_1(x) \\ f_2(x) \\ \vdots \\ f_m(x) \end{pmatrix}$$

The *derivative* of $f(x)$ is denoted Df and it is defined to be an $m \times n$ matrix of partial derivatives:

$$Df[i,j] = \frac{\partial f_i}{\partial x_j} \quad \text{for} \quad 1 \le i \le m,\ 1 \le j \le n$$

Thus, *rows* of Df correspond to output coordinates. Row i of Df is, in fact, Df_i. Thus, we could write

$$Df = \begin{pmatrix} Df_1 \\ Df_2 \\ \vdots \\ Df_m \end{pmatrix}$$

Also, *columns* of Df correspond to input variables. We will write $D_{x_j}f$ for the j-th column of Df. The derivative Df can also be written $D_x f$ if we want to make a clear reference to the variable x. Thus, we could write

$$D_x f = \begin{pmatrix} D_{x_1}f & D_{x_2}f & \cdots & D_{x_n}f \end{pmatrix}$$

We also have a general equation that corresponds to (3.2). Using e_j as we did there, and with a real number t, the quantity $f(p + t \cdot e_j) - f(p)$ computes a change in the j-th coordinate only. In this setting, the difference is a point in $\mathbb{R}^m$ rather than just a number. We still have the limit equation (3.2), except that now the limit involves points in $\mathbb{R}^m$, so that the limit is coordinate by coordinate.

◇ **Problem 58**

Suppose f is an *affine function*; this means that there is an $m \times n$ matrix A and an $m \times 1$ matrix B such that $f(x) = A \cdot x + B$ for all $x \in \mathbb{R}^n$. Show that $Df = A$.

◇ **Problem 59**

Suppose that f is a real-valued *quadratic function*; this means that there is an $n \times n$ matrix A with $A = A^T$, a $1 \times n$ matrix B and a real number c such that $f : \mathbb{R}^n \to \mathbb{R}$ by $f(X) = X^T \cdot A \cdot X + B \cdot X + c$. Show that $Df = 2 \cdot X^T \cdot A + B$. (Hint: use the matrix multiplication formula for write $f(X)$ in terms of the variables $X[j]$.)

◇

◇ **Problem 60**

Let $f : \mathbb{R}^n \to \mathbb{R}$ and let $g : \mathbb{R}^n \to \mathbb{R}$ with Df and Dg defined on all of $\mathbb{R}^n$. Prove the *product rule*: $D(f \cdot g) = Df \cdot g + f \cdot Dg$.

◇

◇ **Problem 61**

Let x, y be variables over $\mathbb{R}^n$ and define $f(x, y) = x + y$. Find Df. (Note: f maps $\mathbb{R}^{2n}$ into $\mathbb{R}^n$, so Df is an $n \times (2n)$ matrix!)

◇

◇ **Problem 62**

Let x, y be variables over $\mathbb{R}^n$ and define $f(x, y) = x \circ y$. Find Df.

◇

Given $f : V \to \mathbb{R}^m$ where $V \subseteq \mathbb{R}^n$ and given $p \in V$, we write $Df(p)$ for the $m \times n$ matrix we get by evaluating each of the function entries in Df at the point p.

◇ **Problem 63**

Define $f(x, y) = x^2 \cdot y - 2 \cdot x^2 + \frac{1}{4} \cdot y^2 - 6 \cdot y$. Find all points $p \in \mathbb{R}^2$ where $Df(p) = \mathbb{O}$.

◇

You know from one variable calculus that not every function has a derivative; furthermore, a function can be differentiable without the derivative being continuous. In several variables, the situation can get quite complicated. The standard definition of differentiable[6] is not even implied by the existence of partial derivatives. For our work, with its applied emphasis, it will be convenient to assume that our functions are C^1 – that each partial derivative $\partial f_i/\partial x_j$ exists and is continuous at each element of the domain of f.

3.1. The Mean Value Theorem. This theorem of calculus obtains an equation
$$f(q) - f(p) = f'(r) \cdot (q - p)$$
where f is a real-valued differentiable function of a real variable, and r is somewhere between the numbers p and q. The analog equation for a function mapping n into m variables looks like this:

$$(3.3) \qquad f(q) - f(p) = \mathbb{D}(Q) \cdot (q - p)$$

where $q, p \in \mathbb{R}^n$ and $\mathbb{D}$ is an $m \times n$ *matrix* of functions and Q is a matrix of points plugged into the functions in $\mathbb{D}$. Note that the dimensions of $\mathbb{D}$ must be $m \times n$, since $q - p$ is $n \times 1$ and $f(q) - f(p)$ is $m \times 1$. We will need to do some work to make equation (3.3) understandable – it is hard to give examples. When we use this equation, we will see that we never care about the actual entries in $\mathbb{D}$ but only that those entries are in some previous specified open set.

We begin with notation: for a set V, we will say that Q is an $m \times n$ *matrix over* V if $Q[i, j] \in V$ for all $1 \le i \le m$ and $1 \le j \le n$. Thus, Q is not a proper matrix in the usual sense; we might think of Q as a table of points. If $V \subseteq \mathbb{R}^k$ and if Q is $m \times n$ over V, then Q can be thought of as having $k \cdot m \cdot n$ real number entries; from that point of view Q is an ordinary vector. We will use that idea to compute $|Q|$ and related quantities.

Let $V \subseteq \mathbb{R}^n$ be open, and let $f : V \to \mathbb{R}^m$ with $Df(q)$ existing for all $q \in V$. Let Q be an $m \times n$ matrix over V. We define the $m \times n$ matrix $\mathbb{D}f(Q)$ by
$$\mathbb{D}f(Q)[i, j] = \frac{\partial f_i}{\partial x_j}(Q[i, j]) \quad \text{for all} \quad 1 \le i \le m, \quad 1 \le j \le n$$

---

[6]You can find the standard definition in [**5**], for instance.

We call $\mathbb{D}f$ the *varied derivative* of f. In using this notation, when the function f is clear from context we will write $\mathbb{D}$ instead of $\mathbb{D}f$. The varied derivative $\mathbb{D}$ maps the $m \times n$ matrices over V into the $m \times n$ matrices (over $\mathbb{R}$). If f is C^1, then the entries of $\mathbb{D}$ are continuous functions on V, and so $\mathbb{D}$ is continuous on its domain by Proposition 3.2. If $q \in V$ and if Q is the $m \times n$ matrix whose every entry is q, then $\mathbb{D}f(Q) = Df(q)$.

We are still assuming that $Df(q)$ exists for all $q \in V$. Let $p \in V$ and get $r > 0$ so that $B(p, r) \subseteq V$. Let $q \in B(p, r)$, and we want to show that there is an $m \times n$ matrix Q over the closed disk $D(p, |q - p|)$ such that (3.3) holds. To do this, we will use the ordinary Mean Value Theorem on each partial derivative of each coordinate function of f. Let f_i be one of f's coordinates, so that $f_i : B(p, r) \to \mathbb{R}$. Let $q \in B(p, r)$. We define the *stairway* from p to q: let $p_0 = p$, and, given p_{j-1} for $1 \le j \le n$, define

$$p_j[k] = \begin{cases} p_{j-1}[k] & \text{if } k \neq j \\ q[j] & \text{if } k = j \end{cases}$$

We can compute that $|p_j - p| \le |q - p|$ for each j, and so $p_j \in D(p, |q - p|)$ for each j.

For instance, with $n = 4$, we have

$$p_0 = p \quad p_1 = (q[1], p[2], p[3], p[4]) \quad p_2 = (q[1], q[2], p[3], p[4])$$
$$p_3 = (q[1], q[2], q[3], p[4]) \quad p_4 = q$$

We will be considering differences such as $f(p_3) - f(p_2)$. Notice that p_3, p_2 differ only at the third coordinate. Write

$$h(x) = f(q[1], q[2], x, p[4]) \quad \text{so that} \quad h'(x) = \frac{\partial f_i}{\partial x_3}(q[1], q[2], x, p[4])$$

Also

$$f_i(p_3) - f_i(p_2) = h(q[3]) - h(p[3])$$

so that the Mean Value Theorem of calculus applies to this difference.

We return to the general situation. Choose j with $1 \le j \le n$. Since p_j and p_{j-1} differ only in the j-th coordinate, the difference $f_i(p_j) - f_i(p_{j-1})$ is susceptible to the ordinary Mean Value Theorem, thinking of f_i as a function of its j-th coordinate only, with derivative $\partial f_i/\partial x_j$. Thus, there is $r_j \in \mathbb{R}^n$ with $r_j[k] = p_j[k] = p_{j-1}[k]$ for all $k \neq j$, and with $r_j[j]$ between $p[j]$ and $q[j]$, and such that

$$f_i(p_j) - f_i(p_{j-1}) = \frac{\partial f_i}{\partial x_j}(r_j) \cdot (q[j] - p[j])$$

Also, notice that $|r_j - p| \leq |p_j - p|$, and so we have $r_j \in D(p, |q - p|)$.

Now since $p_0 = p$ and $p_n = q$, we can write

$$f_i(q) - f_i(p) = \sum_{j=1}^{n} f_i(p_j) - f_i(p_{j-1}) = \sum_{j=1}^{n} \frac{\partial f_i}{\partial x_j}(r_j) \cdot (q[j] - p[j])$$

The r_j were chosen for a particular coordinate function f_i. Now write $Q[i,j] = r_j$ to identify i specifically. Then our equation is this.

$$f_i(q) - f_i(p) = \sum_{j=1}^{n} \frac{\partial f_i}{\partial x_j}(Q[i,j]) \cdot (q[j] - p[j])$$

Recognizing the matrix multiplication formula on the right, this says that the i-th entry of $f(q) - f(p)$ is equal to the i-th entry of $\mathbb{D}(Q) \cdot (q - p)$. We have established (3.3). Here is a statement of the general theorem.

MEAN VALUE THEOREM. *Let $V \subseteq \mathbb{R}^n$ be open, and let $f : V \to \mathbb{R}^m$ with Df defined on that set. Choose $p \in V$ and $r > 0$ so that $B(p,r) \subseteq V$. For each $q \in B(p,r)$ there is an $m \times n$ matrix Q over $D(p, |q - p|)$ such that*

$$f(q) - f(p) = \mathbb{D}(Q) \cdot (q - p)$$

■

A simple but important corollary follows. Recall that the definition C^1 said that the *partial derivatives* were continuous; it did not say that the function was. But it is.

PROPOSITION 3.10. *Let $V \subseteq \mathbb{R}^n$ be open and let $f : V \to \mathbb{R}^m$ be C^1. Then f is continuous on V.*

PROOF. Let $p \in V$, and define P to be the $m \times n$ matrix all of whose entries are p, and then $\mathbb{D}(P) = Df(p)$. Because f is C^1, the function $\mathbb{D}$ is continuous, and so there is $\delta > 0$ such that if Q is $m \times n$ over $B(P, \delta)$, then $|\mathbb{D}(Q) - \mathbb{D}(P)| < 1$. It follows that $|\mathbb{D}(Q)| \leq |\mathbb{D}(P)| + 1$.

Get $r > 0$ with $B(p, r) \subseteq V$ and such that if Q is $m \times n$ over $B(p, r)$, then $Q \in B(P, \delta)$. If $q \in B(p, r)$ then the Mean Value Theorem finds an $m \times n$ matrix Q over $D(p, |q - p|)$ with

$$f(q) - f(p) = \mathbb{D}(Q) \cdot (q - p)$$

We see that Q is $m \times n$ over $B(p, r)$ and so $Q \in B(P, \delta)$, we can estimate

$$
\begin{aligned}
|f(q) - f(p)| &= |\mathbb{D}(Q) \cdot (q - p)| \\
&\leq |\mathbb{D}(Q)| \cdot |q - p| \\
&\leq \left[|\mathbb{D}(P)| + 1 \right] \cdot |q - p|
\end{aligned}
$$

Since f has a Δ-bound at p, we see that it is continuous at p. ∎

We need a converse to the Mean Value Theorem.

PROPOSITION 3.11. *Let $V \subseteq \mathbb{R}^n$ be open, and let $f : V \to \mathbb{R}^m$. Choose $p \in V$ and $r > 0$ so that $B(p, r) \subseteq V$. Suppose that for each $q \in B(p, r)$ there is an $m \times n$ matrix $Q(q)$ such that*

$$f(q) - f(p) = Q(q) \cdot (q - p)$$

and suppose that $Q(q) \to Q(p)$ as $q \to p$. Then $Df(p)$ exists, and $Df(p) = Q(p)$.

PROOF. Let e_j be the j-th column of I_n, for $1 \leq j \leq n$. Let t be a real number with $|t| < r$, and observe that

$$p + t \cdot e_j \in B(p, r)$$

and so we have

$$f(p + t \cdot e_j) - f(p) = Q(p + t \cdot e_j) \cdot t \cdot e_j$$

The product $Q(p + t \cdot e_j) \cdot e_j$ is the j-th column of $Q(p + t \cdot e_j)$, call it Q_j. If $t \neq 0$, we have

$$\frac{f(p + t \cdot e_j) - f(p)}{t} = Q_j$$

Letting $t \to 0$, we have $p + t \cdot e_j \to p$, and so Q_j goes to the j-th column $Q(p)_j$ of $Q(p)$. This proves that $D_{x_j} f(p)$ exists and is equal to $Q(p)_j$. Since this is true of each column, we have $Df(p) = Q(p)$.   ∎

The Mean Value Theorem leads to the Chain Rule.

THE CHAIN RULE. *Suppose that $U \subseteq \mathbb{R}^n$ is open, that $V \subseteq \mathbb{R}^m$ is open, that $f : U \to V$ is C^1, and that $g : V \to \mathbb{R}^k$ is C^1. Then $g(f(x))$ is C^1 and*

$$D(g(f(x))) = Dg(f(x)) \cdot Df(x)$$

PROOF. Let $p \in U$. Since $f(p)$ is an element of the open set V, there is $r' > 0$ such that $B(f(p), r') \subseteq V$. Because f is continuous, there is $r > 0$ so that $B(p, r) \subseteq U$ and $f : B(p, r) \to B(f(p), r')$.

Let $q \in B(p, r)$, and the Mean Value Theorem finds an $m \times n$ matrix $Q(q)$ over $D(p, |q - p|)$ such that

(3.4) $$f(q) - f(p) = \mathbb{D}f(Q(q)) \cdot (q - p)$$

We have $f(q) \in B(f(p), r')$, and the Mean Value Theorem finds $R(q)$ that is $k \times m$ over $D(f(p), |f(q) - f(p)|)$ such that

(3.5) $$g(f(q)) - g(f(p)) = \mathbb{D}g(R(q)) \cdot (f(q) - f(p))$$

Putting together (3.4) and (3.5), we get

(3.6) $$g(f(q)) - g(f(p)) = \mathbb{D}g(R(q)) \cdot \mathbb{D}f(Q(q)) \cdot (q - p)$$

Let P be the $m \times n$ matrix whose every entry is p. We aim to apply Proposition 3.11. The entries of $Q(q)$ are in $D(p, |q - p|)$. As $q \to p$, we see that $Q(q) \to P$. Since f is C^1, the matrix function $\mathbb{D}f$ is continuous, and so $\mathbb{D}f(Q(q)) \to \mathbb{D}f(P) = Df(p)$ as $q \to p$. Proposition 3.10 says that f is continuous, and so as $q \to p$, we have $f(q) \to f(p)$. Since the entries of $R(q)$ are in $D(f(p), |f(q) - f(p)|)$, we see that $R(q)$ approaches the matrix P' all of whose entries are $f(p)$. In other words, $R(q) \to P'$ as $q \to p$. Since g is C^1, the function $\mathbb{D}g$ is continuous, and so $\mathbb{D}g(R(q)) \to \mathbb{D}g(P') = \mathbb{D}g(f(p))$. The upshot:

$$\mathbb{D}g(R(q)) \cdot \mathbb{D}f(Q(q)) \to Dg(f(p)) \cdot Df(p) \quad \text{as} \quad q \to p$$

Proposition 3.11 shows that $g(f(x))$ has the derivative stipulated by this result.

Since Df and Dg are continuous, the formula for $D(g(f(x)))$ shows that it is continuous. Thus, $g(f(x))$ is C^1. ∎

◇ **Problem 64**

Let x, y be variables over $\mathbb{R}^m$ and let $f(x, y) = x + y$. (This was done in a problem on p.82.) Let g, h be C^1 functions that map $\mathbb{R}^n \to \mathbb{R}^m$, and notice that $g(z) + h(z) = f(g(z), h(z))$ for all $z \in \mathbb{R}^n$. Use the formula for Df and the Chain Rule to show that $D(g + h) = Dg + Dh$.
◇

In one-variable calculus an extreme point interior to an interval can occur where the derivative is 0 (or undefined.) Here is the analog in several variables.

INTERIOR EXTREME THEOREM. *Let $U \subseteq \mathbb{R}^n$, and let $f : U \to \mathbb{R}$. Suppose that $p \in U$ and that there is a positive number r such that $B(p, r) \subseteq U$ and f is C^1 on U. Finally, suppose that $f(p)$ is the minimum or maximum of f on U. Then $Df(p) = \mathbb{O}$.*

PROOF. Let $v \in \mathbb{R}^n$ and define $h(t) = p + t \cdot v$ where $t \in \mathbb{R}$. Because $h(0) = p$ and $B(p, r) \subseteq U$, there is[7] $\delta > 0$ such that if $-\delta < t < \delta$, then $h(t) \in B(p, r)$, and so $h : (-\delta, \delta) \to U$. We see that $Dh = v$.

The function $f(h(t))$ maps the open interval $(-\delta, \delta)$ to the real numbers, and its derivative can be computed using the Chain Rule: $Df(h(t)) \cdot Dh(t) = Df(h(t)) \cdot v$. This function has a maximum or minimum when $t = 0$ (when $h(0) = p$). Since $t = 0$ is interior to the open interval $(-\delta, \delta)$, calculus tells us that the derivative of $f(h(t))$ is 0 when $t = 0$. In other words, $Df(p) \cdot v = 0$. Since $Df(p) \cdot v = 0$ for all $v \in \mathbb{R}^n$, we see that $Df(p) = \mathbb{O}$. ∎

_____

[7]If $v = \mathbb{O}$, let $\delta = 1$; if $v \neq \mathbb{O}$, let $\delta = r/|v|$.

3.2. Clumping. It will be convenient to be able to group variables together; we will call such a grouping a *clump*. For instance, suppose that $x \in \mathbb{R}^4$. We might write $x = (y, z)$ where $y = (x_1, x_2, x_3)$ and $z = x_4$. Both y and z are *clumps*. The variable x is also a clump.

Suppose that x is a clump of k variables among the variables of the function f, and that f maps into $\mathbb{R}^m$. Then $D_x f$ stands for the $m \times k$ matrix of partial derivatives with respect to the various variables in x. In other words, $D_x f$ is a collection of columns of Df.

◇ **Problem 65**
Let $f : \mathbb{R}^3 \to \mathbb{R}$ by $f(x, y, z) = 3 \cdot x^2 + x^3 \cdot y - x \cdot z^2$. Let u be the clump (x, z). Find $D_u f$.
◇

Another example. $f : \mathbb{R}^4 \to \mathbb{R}^3$ by

$$f\begin{pmatrix} a \\ b \\ c \\ d \end{pmatrix} = \begin{pmatrix} f_1(a, b, c, d) \\ f_2(a, b, c, d) \\ f_3(a, b, c, d) \end{pmatrix}$$

Suppose we clump $x = (a, b)$ and $y = (c, d)$. Then notice that

$$D_x f = \begin{pmatrix} \partial f_1/\partial a & \partial f_1/\partial b \\ \partial f_2/\partial a & \partial f_2/\partial b \\ \partial f_3/\partial a & \partial f_3/\partial b \end{pmatrix} \quad \text{and} \quad D_y f = \begin{pmatrix} \partial f_1/\partial c & \partial f_1/\partial d \\ \partial f_2/\partial c & \partial f_2/\partial d \\ \partial f_3/\partial c & \partial f_3/\partial d \end{pmatrix}$$

and so we can write Df in blocks:

$$Df = \begin{pmatrix} D_x f & D_y f \end{pmatrix}$$

Now suppose we clump the outputs: let

$$g = \begin{pmatrix} f_1 \\ f_2 \end{pmatrix} \quad \text{so that} \quad f = \begin{pmatrix} g \\ f_3 \end{pmatrix}$$

and now the derivative looks like this.

$$Df = \begin{pmatrix} D_x g & D_y g \\ D_x f_3 & D_y f_3 \end{pmatrix}$$

This is a typical use of clumping: to group inputs or outputs together and take advantage of matrix block notation.

◇ Problem 66

Let x be a variable over $\mathbb{R}^n$ and let y be a variable over $\mathbb{R}^m$. Let $h : \mathbb{R}^n \to \mathbb{R}^m$ be C^1. Let $g : \mathbb{R}^{n+m} \to \mathbb{R}$ be C^1 and clump the inputs to g as x, y. Define $J(x) = g(x, h(x))$. Find a formula for DJ in terms of $D_x g$ and $D_y g$ and Dh.
◇

4. Implicit Curves

We want to motivate a very technical definition that we will need in the study of non-linear problems. Here is a rough description: we imagine standing at a point where some constraint equations hold, and we want to be able to move away from that point while maintaining the equations. We will think about this abstractly and then move to concrete examples.

We imagine m equations in n variables:

$$g_1(x_1, x_2, \ldots, x_n) = 0$$
$$g_2(x_1, x_2, \ldots, x_n) = 0$$
$$\vdots$$
$$g_m(x_1, x_2, \ldots, x_n) = 0$$

We write x for the n variables and $g(x)$ for the m left sides, so that the m equations can be written as a single equation:

$$g(x) = \mathbb{O}_{m \times 1}$$

The function g maps $\mathbb{R}^n$ into $\mathbb{R}^m$.

Now suppose we have a point $p \in \mathbb{R}^n$ such that $g(p) = \mathbb{O}_{m \times 1}$, and suppose there is a curve passing through p that keeps the equation valid. In other words, suppose there is an open interval I in the real numbers and $h : I \to \mathbb{R}^n$ such that $h(0) = p$ and $g(h(t)) = \mathbb{O}$ for all $t \in I$. Such a curve is called a *level curve*. If we assume that g and h are C^1, then we can differentiate the equation $g(h(t)) = \mathbb{O}$. Since the right side is $\mathbb{O}$, its derivative is $\mathbb{O}$! We can apply the Chain Rule to the left side.

$$(3.7) \qquad\qquad Dg(h(t)) \cdot Dh(t) = \mathbb{O}$$

If we plug in $t = 0$, and we remember that $h(0) = p$, this equation becomes

$$(3.8) \qquad\qquad Dg(p) \cdot Dh(0) = \mathbb{O}$$

The $n \times 1$ matrix $Dh(0)$ is the *initial direction* (or initial *velocity*) of the curve h as it moves away from p.

We repeat what we just did: given a level curve $h(t)$ starting at p, moving initially in the direction $Dh(0)$, we have the equations (3.7) and (3.8). We are interested in the converse: given p and a desired initial direction $v \in \mathbb{R}^n$, we ask whether there is a level curve $h(t)$ such that $h(0) = p$ and $Dh(0) = v$. Because we will need (3.8) to be true, we would want $Dg(p) \cdot v = \mathbb{O}$, so we will assume that v has this property.

Here are some examples.

Example 1. Suppose that g is an affine function: $g(x) = A \cdot x + B$ where A is $m \times n$ and B is $m \times 1$. Let $p \in \mathbb{R}^n$ with $g(p) = \mathbb{O}$. We have $Dg = A$, and the equation $Dg(p) \cdot v = \mathbb{O}$ is just that $A \cdot v = \mathbb{O}$. For such a v, define $h(t) = p + t \cdot v$, so that $h : \mathbb{R} \to \mathbb{R}^n$, and compute that $h(0) = p$ and $Dh(0) = v$ and

$$g(h(t)) = A \cdot (p + t \cdot v) + B = A \cdot p + B + t \cdot A \cdot v = \mathbb{O}$$

Thus, there is a level curve in every possible direction.

Example 2. Suppose that $k : \mathbb{R}^n \to \mathbb{R}^m$, and consider a constraint of the form $y = k(x)$ where x has n variables and y has m variables. Let $p_1 \in \mathbb{R}^n$ and $p_2 = k(p_1)$, so that $p = (p_1, p_2)$ satisfies the constraint $y = k(x)$. We show there are level curves in every possible direction.

Indeed, our constraint function can be written $g(x, y) = k(x) - y$; it maps $\mathbb{R}^{n+m}$ to $\mathbb{R}^m$. We have

$$Dg = \begin{bmatrix} Dk & -I_m \end{bmatrix}$$

For $v \in \mathbb{R}^{n+m}$, we write $v = (v_1, v_2)$, clumped as x, y are. Then equation (3.8) looks like this:

$$D_x k(p_1) \cdot v_1 - I_m \cdot v_2 = \mathbb{O}$$

so that $v_2 = D_x k(p_1) \cdot v_1$, with v_1 arbitrary. Given $v_1 \in \mathbb{R}^n$, define $h_1(t) = p_1 + t \cdot v_1$ and $h_2(t) = k(h_1(t))$, so that $h(t) = (h_1(t), h_2(t))$ gives a curve with $h(0) = (p_1, p_2) = p$ that satisfies the constraint. We have

$$Dh(0) = \begin{pmatrix} Dh_1(0) \\ Dh_2(0) \end{pmatrix} = \begin{pmatrix} v_1 \\ D_y k(p_1) \cdot v_1 \end{pmatrix} = \begin{pmatrix} v_1 \\ v_2 \end{pmatrix}$$

and we have the curve we wanted. ∎

Here is a specific example of the type just considered.

Example 2a. The point $(1, 2, -1)$ satisfies the equations $x^2 + y^2 + z^2 - 6 = 0$ and $(x-1)^2 + (y+1)^2 + z^2 - 10 = 0$. Define

$$g(x, y, z) = \begin{pmatrix} x^2 + y^2 + z^2 - 6 \\ (x-1)^2 + (y+1)^2 + z^2 - 10 \end{pmatrix}$$

We solve $g = \mathbb{O}$ for x, y, z by performing a messy, non-linear elimination. The steps we are about to execute are not as important as the form of the answer that results, but here are the steps: replace the second equation in $g = \mathbb{O}$ by the difference between the first equation and the second:

$$x^2 + y^2 + z^2 = 6$$
$$-x + y = 1$$

Substituting $y = x + 1$ into the first equation, we have

$$x^2 + (x+1)^2 + z^2 = 6$$
$$y = x + 1$$

The first equation can be solved for z, remembering that we are starting at $z = -1 < 0$.

$$(3.9) \qquad\qquad z = -\sqrt{6 - x^2 - (x+1)^2}, \quad y = x + 1$$

and we are in the setting of Example 2. We can find level curves starting at $(1, 2, -1)$ in every possible direction. ∎

Not every point has level curves in all directions.

Example 3. The point $(0,0)$ is on the curve $x^2 - y^3 = 0$. We notice that this curve is $y = x^{2/3}$. Also, $D(x^2 - y^3) = \begin{pmatrix} 2x & -3y^2 \end{pmatrix}$. At $(0,0)$, this is the 1×2 zero matrix. Thus, for all $v \in \mathbb{R}^2$, we have $D(x^2 - y^3)(0,0) \cdot v = (0,0)$. Let $v = (1,0)$, and it is easy to see that there is no curve $h(t)$ with $h(0) = (0,0)$ and $h'(0) = (1,0)$ that stays on the graph. Indeed, the tangent line to the graph at $(0,0)$ is vertical, and v is not vertical. ∎

Example 3 exhibits a vector v that satisfies the equation (3.8) but such that there is no curve in that direction.

The technical result we need about level curves is called the Implicit Curve Theorem. We will derive it as a consequence from a very general and useful inverse function theorem.

INVERSE FUNCTION THEOREM. *Let $V \subseteq \mathbb{R}^n$ be open, and let $f : V \to \mathbb{R}^n$ be C^1. Let $p \in V$, and suppose that $Df(p)$ is invertible. Then there are open sets $U \subseteq V$ and $W \subseteq \mathbb{R}^n$ such that*

(a) $p \in U$
(b) f maps U one to one, onto W
(c) f^{-1} is C^1 on W

PROOF. Let $\mathbb{D}$ be the generalized derivative of f, as usual.

Proposition 1.11 says that $\mu(Df(p)) > 0$, since the matrix is invertible; choose $c > 0$ with $c < \mu(Df(p))$. Proposition 1.12 and the continuity of the partial derivatives of f allow us to find $\delta > 0$ so that $D(p, \delta) \subseteq V$ and if $q \in D(p, \delta)^{n \times n}$, then $\mu(\mathbb{D}(q)) \geq c$. Using the Mean Value Theorem, this leads to

$$(3.10) \qquad |f(u) - f(v)| \geq c \cdot |u - v| \quad \text{for all} \quad u, v \in D(p, \delta)$$

It follows that f is one to one on $D(p, \delta)$, for if $u \neq v$ in (3.10), then clearly $f(u) \neq f(v)$.

Define

(3.11)
$$\epsilon = \frac{1}{3} \cdot \delta \cdot c$$

Claim. $B(f(p), \epsilon) \subseteq f(B(p, \delta))$.

Let $y \in B(f(p), \epsilon)$ and define $h : D(p, \delta) \to \mathbb{R}^n$ by $h(x) = |f(x) - y|^2$. (Notice that h is defined on the *closed* disk. We will discard the boundary of that disk momentarily.) Then h is the composite of C^1 functions, and so it is C^1. The domain of h is a closed disk; the Interior Extreme Theorem finds a minimum of h, say it occurs at x.

Suppose, first, that $|x - p| = \delta$. Then, beginning with the reverse triangle inequality,

$$
\begin{aligned}
|y - f(x)| &\geq |f(x) - f(p)| - |y - f(p)| \\
&\geq c \cdot |x - p| - |y - f(p)| & \text{by (3.10)} \\
&\geq c \cdot \delta - \frac{1}{3} \cdot \delta \cdot c & \text{by (3.11) and } y \in D(f(p), \epsilon) \\
&= \frac{2}{3} \cdot c \cdot \delta > \frac{1}{3} \cdot \delta \cdot c > |y - f(p)|
\end{aligned}
$$

We see that $f(p)$ is closer to y than $f(x)$. This contradicts that $h(x)$ is the minimum of h, and it shows that $|x - p| < \delta$; in other words, $x \in B(p, \delta)$, the *open* disk.

We see that $h(x)$ is the minimum of h on the open disk $B(p, \delta)$, and the Interior Extreme Theorem shows that $Dh(x) = 0$. Compute

$$\mathbb{O} = Dh(x) = 2 \cdot (f(x) - y)^T \cdot Df(x)$$

Since $x \in B(p, \delta)$, the matrix $Df(x)$ is invertible, and so $\mathbb{O} = (f(x) - y)^T$, which is $y = f(x)$. This proves the Claim. ∎

Define

$$U = \left\{ x \in B(p, \delta) \mid f(x) \in B(f(p), \epsilon) \right\}$$

and since f is continuous, the set U is open. The Claim shows that f maps U onto $B(f(p), \epsilon)$, and we know that f is one to one. Thus, $f^{-1} : B(f(p), \epsilon) \to U$ is one to one and onto, as well.

Claim. The function f^{-1} on $B(f(p), \epsilon)$ is continuous.

For let $y \in B(f(p), \epsilon)$ and let $\beta > 0$. Apply what we have already done to the function

$$f : B(p, \delta) \cap B(f^{-1}(y), \beta) \to \mathbb{R}^n$$

There is an open set $W \subseteq \mathbb{R}^n$ with $y \in W$ and with f^{-1} mapping W onto an open subset of $B(p, \delta) \cap B(f^{-1}(y), \beta)$. That does it. ∎

Claim. The function f^{-1} on $B(f(p), \epsilon)$ is C^1.

Indeed, let $y \in B(f(p), \epsilon)$, and define $x = f^{-1}(x)$. Get a disk $B(x, r) \subseteq B(p, \epsilon)$. For each $u \in B(x, r)$, define $v = f(u)$. The Mean Value Theorem finds an $n \times n$ matrix Q over $B(p, \delta)$ such that

$$f(u) - f(x) = \mathbb{D}(Q) \cdot (u - x) \quad \text{which is} \quad v - y = \mathbb{D}(Q) \cdot (f(v) - f(y))$$

We know that $\mathbb{D}(Q)$ is invertible, and so

$$\mathbb{D}(Q)^{-1} \cdot (v - y) = f(v) - f(y)$$

Let $v \to y$. Since f^{-1} is continuous, we have $u \to x$. The Mean Value Theorem then says that $\mathbb{D}(Q) \to Df(x)$.

Then $\mathbb{D}(q(v)) \to Df(f^{-1}(y)) = Df(x)$ by the Mean Value Theorem. Proposition 3.6 says that taking the matrix inverse is continuous, and we see that $\mathbb{D}(Q)^{-1}$ is a continuous function of v. Proposition 3.11 tells us that f^{-1} has a continuous derivative at y. ∎

In the context of the Inverse Function Theorem, if $q = f(p)$, then we see that

$$Df^{-1}(q) = Df(p)^{-1}$$

This is the general version of the formula in one variables

$$\frac{dy}{dx} \cdot \frac{dx}{dy} = 1$$

that governs inverse functions in that case.

IMPLICIT CURVE THEOREM. *Let m, n be positive integers and clump the elements of $\mathbb{R}^{n+m}$ as (x, y) with $x \in \mathbb{R}^n$ and $y \in \mathbb{R}^m$. Let V be an open subset of $\mathbb{R}^{n+m}$; let $g : V \to \mathbb{R}^m$ be C^1. Let $p \in V$ with $g(p) = \mathbb{0}_{m \times 1}$. Suppose* *that $D_y g(p)$ is invertible. Let $v \in \mathbb{R}^{n+m}$ with $Dg(p) \cdot v = \mathbb{O}$. Then there is a positive number δ and $h : [-\delta, \delta] \to V$ such that*

(a) H is C^1
(b) $h(0) = p$
(c) $g(h(t)) = \mathbb{O}$ for all $t \in [-\delta, \delta]$
(d) $Dh(0) = v$

PROOF. Clump $v = (v_x, v_y)$, and the equation $Dg(p) \cdot v = \mathbb{O}$ is this.

$$D_x g(p) \cdot v_x + D_y g(p) \cdot v_y = \mathbb{O}$$

Since $D_y g(p)$ is invertible, we have

$$(3.12) \qquad\qquad v_y = -D_y g(p)^{-1} \cdot D_x g(p)$$

Define $f : V \times \mathbb{R}^m \to \mathbb{R}^{n+m}$ by

$$f(x, y) = \begin{bmatrix} x & g(x, y) \end{bmatrix}$$

Then

$$Df = \begin{bmatrix} I_n & \mathbb{O} \\ D_x g & D_y g \end{bmatrix}$$

and f is C^1. The derivative is invertible at p, and so the Inverse Function Theorem finds an open subset U of $V \times \mathbb{R}^m$ and W of $\mathbb{R}^{n+m}$. with $p \in U$ and such that f^{-1} maps W onto U. Since $f(p) = (p_x, \mathbb{O})$, we see that $f^{-1}(p_x, \mathbb{O}) = p$. We know that

$$Df^{-1}(p_x, \mathbb{O}) = Df(p)^{-1}$$

and so, we see that

$$D_x f^{-1}(p_x, \mathbb{O}) = \begin{bmatrix} I_n \\ D_y g^{-1}(p) \cdot D_x g(p) \end{bmatrix}$$

and so

$$D_x f^{-1}(p_x, \mathbb{O}) \cdot v_x = \begin{bmatrix} I_n \\ D_y g^{-1}(p) \cdot D_x g(p) \end{bmatrix} \cdot v_x$$

$$= \begin{bmatrix} v_x \\ D_y g^{-1}(p) \cdot D_x g(p) \cdot v_x \end{bmatrix}$$

$$= \begin{bmatrix} v_x \\ v_y \end{bmatrix} \qquad \text{using (3.12)}$$

This shows that

(3.13) $$D_x f^{-1}(p_x, \mathbb{O}) \cdot v_x = v$$

We intend to define $h(t) = f^{-1}(p_x + t \cdot v_x, \mathbb{O})$ for $t \in \mathbb{R}$ and where the zero vector is in $\mathbb{R}^m$. Since $f(p) = (p_x, \mathbb{O})$, we have $(p_x, \mathbb{O}) \in W$, and there is $\delta \geq 0$ such that h is defined for $t \in [-\delta, \delta]$. The definition of h shows that it is C^1, and that

$$f(h(t)) = (p_x + t \cdot v_x, \mathbb{O}) \quad \text{so that} \quad h(t)_x = p_x + t \cdot v_x \quad \text{and} \quad g(h(t)) = \mathbb{O}$$

Since $f(p) = (p_x, g(p)) = (p_x, \mathbb{O})$, we see that $h(0) = f^{-1}(p_x, \mathbb{O}) = p$.

Compute

$$Dh(0) = D_x f^{-1}(p_x, \mathbb{O}) \cdot v_x = v$$

using (3.13). ∎

We call attention to the formula $h(t)_x = p_x + t \cdot v_x$ that occurred in the proof. We will make use of that detail later.

CHAPTER 4

Non-linear Optimization

1. Introduction

It is a major departure to tamper with the special hypotheses that define a linear program. It might help to begin with the terminology that transfers to the general situation.

The quantity being minimized is still called the objective, although it may be a complicated function of the problem variables. In contrast to linear programming, we generally do not require that the problem variables be non-negative. If we wish to impose that condition, we do so via the constraints described in the next paragraph.

Constraints of a general optimization problem can be put into primal form or equation form as in the case of a linear program, but we are more interested in a mixture of the two forms called a *constraint complex*. A constraint complex involves some number m of constraints over some number n of variables. The complex consists of an open subset V of $\mathbb{R}^n$, a C^1 function $g : V \to \mathbb{R}^m$, and a vector of m symbols, each of which is "=" or "≤." For each integer i with $1 \le i \le m$, the i-th entry $g_i(x)$ of g, and the i-th symbol (= or ≤) define a single constraint with right side 0.

The *feasible set* for this constraint complex is the set of $v \in V$ such that, for each i with $1 \le i \le m$ we have whichever of $g_i(v) \le 0$ or $g_i(v) = 0$ is implied by the choice of the i-th constraint. The vector v is called a *feasible vector*.

Beginning with the simplest example, the constraint $1 \le x \le 4$ on a single variable x can be written as $1 - x \le 0$ and $x - 4 \le 0$, so that $g(x)$ maps $\mathbb{R}$ to $\mathbb{R}^2$.

◇ **Problem 67**

Identify the constraint complex if the constraints are these: $x^2 + y^2 + z^2 \leq 4$, $x \geq 0$, $y \geq x^2 + 1$, $x^2 + y \cdot z = 4$.

◇

Our constraints have involved equations and non-strict inequalities. What about *strict* inequalities? Suppose that the domain of our problem is the open subset V of $\mathbb{R}^n$. A strict inequality such as $x[1] < 5$ can be incorporated into the domain: define V' to be the set of $x \in V$ such that $x[1] < 5$. The set V' is open, and we no longer need $x[1] < 5$ as a constraint, since it is implicit in the definition of V'.

◇ **Problem 68**

For real variables x, y, define an open set V and a constraint complex to describe the following: $2 < x < 7$ and $-3 < y \leq 9$ and $x^2 + y \geq 1$.

◇

Given a constraint complex: $g : V \rightarrow \mathbb{R}^m$ with $V \subseteq \mathbb{R}^n$ and with a vector of equality/inequality choices, we might consider an objective $f : V \rightarrow \mathbb{R}$ and ask for the minimum of f over the feasible set of the complex. The possibilities for existence of a solution transfer from the linear situation: a problem is *infeasible* if there are no feasible vectors; a problem is *feasible* otherwise. The objective in a feasible problem may or may not have a minimum over the set of feasible vectors. If a minimum exists, it will be unique, although there may be several points at which the objective takes on that minimum value. A point where the minimum occurs is a *solution*.

Here is a typically convoluted problem to get us started.

◇ **Problem 69**

Solve this problem: maximize a such that the maximum of $x^3 - 27 \cdot x - 30$ on $x \leq a$ is 24.

◇

The Interior Extreme Theorem was stated on p.88. Here is an application of that fact on *unconstrained curve fitting*. For instance, suppose we are given data points (x_j, y_j) for $1 \leq j \leq n$ in the plane and we want to find their *regression line*, the *line of best fit* to them. You probably know how to do this: given a line $y = a \cdot x + b$, we measure the fit of the line to the data, the *squares error*,

$$E(a, b) = \sum_{j=1}^{n} (a \cdot x_j + b - y_j)^2$$

The regression line is the line whose squares error is minimum. In the following problem, we *assume* there is a minimum to the function $E(a, b)$; later we will prove that the minimum exists.

◇ **Problem 70**

Let (x_j, y_j) for $1 \leq j \leq n$ be given points in the plane, and let E be the squares error. Assuming that E has a minimum at some point $(a, b) \in \mathbb{R}^2$, find (a, b).

◇

Curve fitting can apply to curves other than lines.

◇ **Problem 71**

★ Measure the fit of data points (x_j, y_j) for $1 \leq j \leq n$ to the cubic curve $a \cdot x^3 + b \cdot x^2 + c \cdot x + d$ by the squares error

$$E(a, b, c, d) = \sum_{j=1}^{n} \left(a \cdot x_j^3 + b \cdot x_j^2 + c \cdot x_j + d - y_j \right)^2$$

Find (a, b, c, d) to minimize E where the data points are these: $(0, 0)$, $(1, 1)$, $(2, 0)$, $(3, 4)$, $(4, -1)$.

◇

The general linear curve fitting problem tries to find $n \times 1$ matrix X to minimize $|A \cdot X - B|^2$, where A is a given $m \times n$ matrix and B is given $m \times 1$. One usually imagines that A, B are determined by data of some sort.

Problem 72

Let A be $m \times n$ and let B be $m \times 1$. Assume that the function

$$E(X) = |A \cdot X - B|^2$$

has a minimum for $X \in \mathbb{R}^n$. Use the Interior Extreme Theorem to write
down a system of linear equations that must be satisfied by the value of X
where the minimum occurs.[1] (Hint: use the transpose to write $E(X)$ as a
matrix product and expand; a problem on p.82 of Chapter 3 gave a formula
for the derivative of such a function.)
$\Diamond$

 Problem 73

Let Y be a solution to the equations of the previous problem. Let $X \in \mathbb{R}^n$,
and show that $|A \cdot X - B|^2 - |A \cdot Y - B|^2 = |A \cdot (X - Y)|^2$. (Note: this
shows that Y is a solution to the minimization problem.)
$\Diamond$

You know from experience that non-linear equations (constraints) may be
algebraically intractable, and so we cannot expect to be systematic in de-
scribing the set of feasible vectors. But although there is no analog of the
elimination algorithm for solving a set of arbitrary constraints, the hypothesis
given in the next section will support the notion of *free variables* and *basic
variables* identified in the multivariate calculus, and this structure will lead to
the Kuhn-Tucker conditions alluded to previously.

We mention that the approximate, numerical solution of optimization problems is an ongoing area of research. We will find many but not all such problems susceptible to the standard software, and you are cautioned not to be too
cavalier about solutions obtained on a computer. There are more possible nu-
ances in non-linear problems than in linear problems, and apparent numerical
solutions can hide significant pathologies.

---

[1] It is a fact that the linear equations here must have a solution. The equations are
called the *normal equations*.

2. Full Row Rank

Suppose we have a constraint complex consisting of $V \subseteq \mathbb{R}^n$ and a C^1 function $g : V \to \mathbb{R}^m$ and choices of equation/inequality. Let p be a feasible vector for the complex. We are ready to describe a hypothesis that will ensure certain conditions when the minimum of an objective function occurs at p. We begin by collecting the outputs $g_i(p)$ that are equal to 0. Say there are k of them (it could be that $k = 0$). Renumber the g_i to put these outputs first. Define

$$G(x) = \begin{bmatrix} g_1(x) \\ g_2(x) \\ \vdots \\ g_k(x) \end{bmatrix} \quad \text{and} \quad H(x) = \begin{bmatrix} g_{k+1}(x) \\ \vdots \\ g_m(x) \end{bmatrix}$$

Then we have $G(p) = \mathbb{O}$, and $H(p) < \mathbb{O}$. The constraints in H must be of inequality type, since p is feasible. We call G the *zero part of* g, p, and H is the *non-zero part of* g, p.

We say that g, p has *full row rank* if $DG(p)$ has rank k. Notice that the matrix $DG(p)$ is $k \times n$, and so full row rank says that the rank of $DG(p)$ is equal to its number of rows. There is a special case: if $g_i(p) < 0$ for all i, so that G is *empty*, then we say that g, p has full row rank in that case, as well.

Let's begin with some practice identifying this condition.

◇ **Problem 74**
In each case, find any points on the constraints that do not have full row rank.

(a) $x^2 \leq y \leq 4$
(b) $y^2 - x^2 = 0$ and $-1 \leq x \leq 1$
(c) $y \geq x^2$ and $y \leq x + 2$
(d) For a positive constraint c,

$$\sum_{j=1}^{n} x_j^2 \leq c \quad \text{and} \quad x_j \geq 0 \quad \text{for each} \quad j$$

3. The Kuhn-Tucker Conditions

Here is our main hypothesis; it will be in force for the rest of this chapter, although we will mention it explicitly each time it is used. When you apply the techniques we are about to discuss, you need to be sure that this hypothesis holds. Notice in particular that we are *assuming* that the objective f has a minimum.

HYPOTHESIS 4.1. *Suppose we have a constraint complex with $g : V \to \mathbb{R}^m$, where g is C^1. Let $f : V \to \mathbb{R}$ be C^1. Suppose that p is a feasible vector and that g, p has full row rank. Let $G(x)$ be the zero part of $g(x)$ for p, and let $H(x)$ be the non-zero part. Suppose that the minimum of f on the feasible vectors occurs at p.*

Recall that Proposition 2.6b has the Lagrange Multiplier equation for a linear program.

The Lagrange Multiplier   There is a $1 \times m$ matrix λ with

$$Df(p) = \lambda \cdot Dg(p)$$

The entries of the matrix λ are usually called *Lagrange Multipliers*, or we can refer to λ as a whole as *the* Lagrange Multiplier. We will see that λ does not have to be unique; its variety will become clear in the course of proving that it exists.

PROOF. Let U be the set of $x \in V$ such that $H(x) < \mathbb{O}$. Since V is an open set, the set U is also open. Let F be the set of $x \in U$ such that $G(x) = \mathbb{O}$. We claim that the elements of F are feasible vectors (for the problem indicated in Hypothesis 4.1). Indeed, if $x \in F$, then $x \in U \subseteq V$, and so $G(x) = \mathbb{O}$ and $H(x) < \mathbb{O}$.

We see that $p \in F$, and so since p is a solution to the original problem, we see that p is the solution to this problem:

Problem A   Minimize $f(x)$ over $x \in F$.

We will use Lemma 1.7 to find a $1 \times k$ matrix λ such that $Df(p) = \lambda \cdot DG(p)$. To do this, let $v \in \mathbb{R}^n$ with $DG(p) \cdot v = \mathbb{O}$. The hypothesis that g, p has full row rank shows that $DG(p)$ has rank k. The Implicit Curve Theorem finds $\delta > 0$ and $h : [-\delta, \delta] \to U$ such that $h(0) = p$ and $Dh(p) = v$ and $G(h(t)) = \mathbb{O}$ for all $t \in [-\delta, \delta]$. For each t, we have $h(t) \in U$, and the definition of the set F shows that $h(t) \in F$. In particular, $h(t)$ is feasible for Problem A. Then the function $f(h(t))$ maps $[-\delta, \delta]$ into $\mathbb{R}$ and has minimum at $t = 0$, since $h(0) = p$. The Interior Extreme Theorem says that $D(f(h(t)) = \mathbb{O}$ when $t = 0$. In other words,

$$\mathbb{O} = Df(h(0)) \cdot Dh(0) = Df(p) \cdot v$$

We have proved that if $DG(p) \cdot v = \mathbb{O}$, then $Df(p) \cdot v = \mathbb{O}$. By Lemma 1.7, there is a $1 \times k$ matrix λ such that $Df(p) = \lambda \cdot DG(p)$.

Now notice that

$$Df(p) = \lambda \cdot DG(p) = \begin{bmatrix} \lambda & \mathbb{O} \end{bmatrix} \cdot \begin{bmatrix} DG(p) \\ DH(p) \end{bmatrix}$$

The matrix on the far right is $Dg(p)$. We have the Lagrange Multiplier. ■

The way we constructed the Lagrange Multiplier, we had $\lambda[i] = 0$ for constraints such that $g_i(p) < 0$. This condition is called *Complementary Slackness*, and we will always enforce it. Proposition 2.6c gives complementary slackness in the case of a linear program.

In the case that G is empty (that $g = H$), we have $U = F$, so that F is an open set. Since p is a solution to Problem A, the Interior Extreme Theorem shows that $Df(p) = \mathbb{O}$. In this case the Lagrange Multiplier is $\mathbb{O}$, consistent with Complimentary Slackness.

Complementary Slackness. Assume Hypothesis 4.1. In the Lagrange Multiplier λ, we can have $\lambda[i] = 0$ whenever $g_i(p) < 0$.

◇ Problem 75

Let $v \approx 8$ be an unknown positive constant. For the problem: minimize $-4x - y^2$ with $x^2 + y^2 \leq v$, find all solutions to the the Lagrange Multiplier equation, and find a solution to the problem.

◇

Problem 76

Show that the Lagrange multiplier equations have a unique solution in the problem: minimize y such that $y - x^3 = 0$, but that this solution does not give a minimum. Why not?

◇

 Problem 77

Consider the problem: minimize y such that $y^2 - 2x^2y + x^4 = 0$. Show that the Lagrange multiplier equations do not hold at the minimum. What's wrong?

◇

Perturbation Curves and Shadow Prices   An inequality of the form $g(x) \leq b$, where b is a constant, can be converted to $g(x) - b \leq 0$. Notice that $D(g(x) - b) = Dg(x)$, and so, from the point of view of Lagrange Multipliers, it doesn't matter whether we leave the b on the right side or move it to the left.

Consider, then, a single constraint of the form $g(x) \leq b$ and suppose that p is a specific point with $g(p) = b$. As in the LP case, we call b a *resource bound*, and we want to consider the effect on an objective function of perturbing (changing) b. To do this, we replace the constraint $g(x) \leq b$ by the constraint $g(x) - t = b$, where t is a (new) real variable; in fact, t is a *slack variable*. If x becomes a function of t with $x(0) = p$, so that our equation will be $g(x(t)) - t = b$, and if we have an objective function f, then we can consider $f(x(t))$. The *shadow price* of this perturbation is the derivative of $f(x(t))$ at $t = 0$. We will show that under Hypothesis 4.1, the curve $x(t)$ necessarily exists, and no matter which such curve $x(t)$ we use, we get the same shadow price.

Recall that Proposition 2.6b and Proposition 2.9 found the shadow price for a constraint in an LP; the shadow price is non-positive in that situation. In the general case, the sign of the shadow price may or may not be determined.

As before, assume that Hypothesis 4.1 holds. Recall that Dg is $m \times n$. Choose a single constraint $g_i \leq b$. If $g_i(p) \neq b$, then we can perturb the right side of the constraint without affecting p. Therefore, we say that the shadow price is 0 in this case. Since Complementary Slackness says that the coordinate $\lambda[i]$ of the Lagrange Multiplier is also 0, the shadow price agrees with the Lagrange Multiplier.

Now assume that $g_i(p) = b_i$. Using the zero and non-zero parts G, H as before, the constraint g_i is part of G. We assume that $i = 1$ and $G_1 = g_1$. Assume that G has k coordinates, and let E be column 1 of I_k. Because the matrix $DG(p)$ has rank k, there is a $n \times 1$ matrix v such that

$$(4.1) \qquad DG(p) \cdot v = E$$

Now consider the function

$$K(x,t) = G(x) - t \cdot E \quad \text{so that} \quad DK(x,t) = \begin{bmatrix} DG(x) & -E \end{bmatrix}$$

Then $K(p,0) = b$ and $DK(p,0)$ has rank k, since $DG(p)$ has rank k. This means that the rank of $DK(p,0)$ is its number of rows. Also, (4.1) shows that

$$DK(p,0) \cdot \begin{bmatrix} v \\ 1 \end{bmatrix} = DG(p) \cdot v - E = \mathbb{O}$$

Notice that the t variable is not involved in the invertible part of the derivative. By the Implicit Curve Theorem there is $\delta > 0$ and a C^1 function $h : (-\delta, \delta) \to V$ such that $h(0) = p$ $Dh(0) = v$ and

$$K(h(t), t) = b \quad \text{for all} \quad t$$

Write $x = h(t)$, and we have $K(x,t) = G(h(t)) - t \cdot E$ for all t. In other words $G(h(t)) - t \cdot E = b$ for all t. We can drop K and concentrate on $h(t)$ – a curve which accomplishes perturbation.

Recall the Lagrange equation $Df(p) = \lambda \cdot DG(p)$. We are now in a position to compute a shadow price:

$$D\big(f(h(t))\big)\big|_{t=0} = Df(p) \cdot Dh(0)$$
$$= \lambda \cdot DG(p) \cdot Dh(0)$$
$$= \lambda \cdot DG(p) \cdot v$$
$$= \lambda \cdot E \qquad\qquad \text{using (4.1)}$$
$$= \lambda[1]$$

Recall that g_1 was the constraint under perturbation. The shadow price was the Lagrange Multiplier for the constraint.

Shadow Price. Assume Hypothesis 4.1. When Complementary Slackness is enforced, the i-th coordinate of the Lagrange Multiplier is the shadow price of the i-th constraint.

◇ **Problem 78**

Consider the problem: maximize $Z = (y - x)^2$ such that $x \geq 0$ and $y \geq 0$, and $x + y \leq 1$.

(a) Find the Lagrange multipliers at each of the two solutions: $(1, 0)$ and $(0, 1)$.

(b) At $(1, 0)$, if we perturb $y \geq 0$ to $y \geq \delta$, where $\delta < 0$, show that the Lagrange multiplier correctly predicts the rate of change of the maximum.

(c) At $(0, 1)$, if we perturb $y \geq 0$ to $y \geq \delta$, where $\delta < 0$, show that the Lagrange multiplier does not correctly predict what happens to the maximum value of Z.

There is one further condition that we want; it is somewhat technical, but it does occur often. Restart with a particular constraint, say it's the first constraint, and assume that the constraint is an inequality: $g_1(x) \leq 0$. Get the perturbation curve $h(t)$ as above, with $h(0) = p$ and $Dh(0) = v$ such that

$$DG(p) \cdot v = E$$

We also have $G(h(t)) = t \cdot E$ for each t in the domain of h.

We claim that $h(t)$ is feasible for the constraint complex when $t \leq 0$. For a constraint g_i in G, if $i > 1$, then $g_i(h(t)) = 0$ for all t. And for $i = 1$, we have $g_1(h(t)) = t$, and this gives $g_1(h(t)) \leq 0$ when $t \leq 0$. Recall that this first constraint was $g_1(x) \leq 0$, and now $h(t)$ is seen to be feasible.

Next we claim that the derivative of $f(h(t))$ at $t = 0$ is non-positive. For if it were positive, there would be a small negative value of t with $f(h(t)) < f(h(0)) = f(p)$, and this would contradict the minimality of $f(p)$ over the feasible points. In other words, the shadow price $\lambda[1]$ is non-positive. This is the *sign condition*.

Sign Condition. Assume Hypothesis 4.1. If the i-th constraint is $g_i(x) \leq 0$, and if $g_i(p) = 0$, then every Lagrange Multiplier λ satisfies $\lambda[i] \leq 0$.

Proposition 2.6a and Proposition 2.9 show that the Lagrange Multipliers for a linear program in primal form are non-positive.

We summarize the conditions we have established. They were introduced in the paper [**2**, pp. 481-492].

Kuhn-Tucker Conditions   Assume Hypothesis 4.1.

  (1) The Lagrange multiplier equations: there is λ such that $Df(p) = \lambda \cdot Dg(p)$.
  (2) Complementary slackness: if $g_j(p) \neq 0$, then $\lambda[j] = 0$.
  (3) The Sign Conditions: if the j-th constraint is $g_j(x) \leq 0$, then $\lambda[j] \leq 0$.

Let's speak about an approach to a general optimization problem. Remember that the Kuhn-Tucker conditions require the existence of a solution and that it occur at a full row rank point. If the constraints define a closed and bounded set, then the Extreme Value Theorem guarantees the existence of both a maximum and minimum. A constraint complex defines a closed feasible set; if the set is not bounded, we might consider imposing a bound, so that the extremes exist. Of course, we would have to figure out what happens when we relax that bound.

Once we are willing to accept the existence of the solution we are seeking, we consider sets of constraints, in turn, as possible zero part; we try to solve the Kuhn-Tucker Conditions, looking also for points that do not have full row rank. We will work examples in class.

◇ Problem 79

In the problem: minimize $f(x)$ such that $g(x) \leq 0$, where $x \in \mathbb{R}$, change $g(x) \leq 0$ to $g(x) + y^2 = 0$ where y is a slack variable. Show that the Lagrange multiplier equation for "minimize $f(x)$ such that $g(x) + y^2 = 0$" includes the complementary slackness equation for the original constraint $g(x) \leq 0$.

◇

◇ Problem 80

Find the maximum and minimum volume of a cylindrical can whose surface area (including the top and the bottom) is 12.

◇

◇ Problem 81

Find the point farthest from $(1, 3, -1)$ such that $x^2 + y^2 + z^2 \leq 11$ and $x - y + z \leq 3$. What happens to the maximum distance if the 11 on the right side of the inequality is perturbed?

◇

◇ Problem 82

Find the maximum and minimum of $6x^3 - 6xy + y^2$ where $x^2 \leq y \leq 12$.

◇

◇ Problem 83

Find the minimum of $x^2 - 2x + y^2 - 4y$ such that $y - x = 1$, $x + y \leq 2$, $x \geq 0$, and $y \geq 0$.

◇

◇ Problem 84

Find the minimum of $-2x - y$ such that $x^2 + y^2 \leq 25$ and $x - y \leq 1$.

◇

4. Applications

We need a linear algebra fact.

◇ **Problem 85**

Let A be an $m \times n$ matrix of rank m. Then $A \cdot A^T$ is invertible. (Hint: if not, then there is a non-zero $m \times 1$ matrix v such that $A \cdot A^T \cdot v = \mathbb{O}$. Show that $|A^T \cdot v| = \mathbb{O}$, and get a contradiction.)

◇

Here is a geometric application.

◇ **Problem 86**

Let A be an $m \times n$ matrix of rank m. If B is $m \times 1$, then the equation $A \cdot X = B$ is consistent. The vector v is the *least effort solution* to this equation if $|v|$ is minimal. Show that $v = A^T \cdot (A \cdot A^T)^{-1} \cdot B$. (Hint: show that the Kuhn-Tucker conditions have to hold, and v is the only solution to those conditions.)

◇

4.1. Inequalities. There are several famous inequalities that can be derived by solving optimization problems.

◇ **Problem 87**

Find the maximum of $x_1 \cdot x_2 \cdots x_n$ such that $\sum_{j=1}^{n} x_j \leq 1$ and $x_j \geq 0$ for $1 \leq j \leq n$.

◇

COROLLARY 4.2. *Let $x_1, \ldots, x_n$ be non-negative real numbers. Then their geometric mean of is no greater than their arithmetic mean. In other words,*

$$\left(\prod_{j=1}^{n} x_j \right)^{1/n} \leq \frac{1}{n} \cdot \sum_{j=1}^{n} x_j$$

PROOF. The result is trivial if all the x_j are 0. Otherwise, let c be the sum of the x_j, so that $c > 0$, and define $y_j = x_j/c$. The sum of the y_j is 1, and so their maximum product is computed by the previous problem. Call this maximum P. We have

$$\prod_{j=1}^{n} y_j \leq P$$

When we write both sides of this inequality in terms of the x_j, we will see the desired inequality between the geometric mean and arithmetic mean. ∎

The next problem establishes *Young's inequality*.

◇ **Problem 88**

Let p, q be positive numbers with $\frac{1}{p} + \frac{1}{q} = 1$. Show that 0 is the minimum of

$$Z = \frac{x^p}{p} + \frac{y^q}{q} - x \cdot y \quad \text{where} \quad x, y \geq 0$$

(Note: Young's inequality is that $Z \geq 0$.)
◇

◇ **Problem 89**

Find the maximum and minimum of

$$\sum_{j=1}^{n} x_j \cdot y_j \quad \text{such that} \quad \sum_{j=1}^{n} x_j^2 \leq 1, \quad \sum_{j=1}^{n} y_j^2 \leq 1$$

◇

This last problem provides another proof of the Cauchy-Schwarz inequality – Proposition 1.8. For let $x, y \in \mathbb{R}^n$. If $x = \mathbb{O}$ or $y = \mathbb{O}$, the inequality is trivial. Otherwise, replace x by $x/|x|$ and y by $y/|y|$ and we have vectors which are feasible for the previous problem. When we write the maximum from the problem in terms of x, y, we will see the Cauchy-Schwarz inequality.

4.2. Production Theory. In this subsection and the next we give two representative applications to economics. See [**6**], or any other book on mathematical economics, for many more. We imagine $v \in \mathbb{R}^m$, an *input vector* representing amounts of various inputs to a production product, and we have $x \in \mathbb{R}^n$, an *output vector* of quantities produced. We regard v, x as the problem variables. We have a C^1 function $F : \mathbb{R}^{m+n} \to \mathbb{R}^k$, where $F(v, x) = \mathbb{O}$ represents the feasibility of producing outputs x from inputs v. A very simple situation: let $f(x)$ be the inputs necessary to producing x outputs, and then we can have $F(v, x) = v - f(x)$. (We have $n = k$ in this case.)

If the unit price of quantity x_j is p_j, and if the unit cost of input v_i is q_i, for all relevant j, i, then the net revenue from production is $R = p \circ x - q \circ v$. The generic production problem is to maximize R subject to $F(v, x) \geq 0$ and $v \geq \mathbb{O}$ and $x \geq \mathbb{O}$.

It seems realistic to assume (1), that input determines output, and (2) that the inputs have upper bounds. We would expect the feasible vectors to define a closed set, and the Extreme Value Theorem tells us that R has a maximum and a minimum on the feasible set.

Here is an example: we produce four outputs x_i from seven inputs $I[j]$. Here are the functions that show the price of each output, given its level of production.

$$p_1 = \frac{5000}{x_1^2 + 100} \qquad\qquad p_2 = \frac{2000}{x_2^2 + 25}$$

$$p_3 = \left(1 - \frac{x_4}{30}\right) \cdot \frac{15000}{x_3^2 + 400} \qquad\qquad p_4 = \frac{3000}{x_4^2 + 36}$$

In the following table, the first four rows show, for each output, the number of units of each input needed to make one unit of that output. The *cost* row shows the unit cost of each input. The *available* row shows how much of each input is available.

	I[1]	I[2]	I[3]	I[4]	I[5]	I[6]	I[7]
output 1:	0	1	2	1	0	0	3
output 2:	2	0	1	0	1	2	4
output 3:	1	1	2	1	1	0	3
output 4:	3	0	2	0	0	0	3
cost:	2	5	3	2	1	4	6
available:	20	6	23	8	6	5	30

 Problem 90

★ Determine how much of each output to produce, and at what price, to maximize net revenue. What would it be worth to us to increase the available supply of input 2 by 1? Of input 7 by 1?

◇

4.3. Activity Analysis. The n activities of a business involve m commodities. The commodities may either be consumed (e.g. electricity used to light the building) or produced (e.g. a finished product for sale). Let A be an $m \times n$ matrix where $A[i, j]$ gives the amount of commodity i involved in one unit of activity j. If $A[i, j] < 0$, then commodity i is consumed by activity j; if $A[i, j] > 0$, then commodity i is produced by activity j.

If $x \in \mathbb{R}^n$ is the activity vector, then $y = A \cdot x$ is the resulting commodity vector. Notice that the entries in y may be negative, indicating a net use of the commodity, and those entries may be positive, indicating net production. We would have $x \geq \mathbb{O}$.

We imagine bounds on the availability of commodities that are consumed. Let R be $m \times 1$ where $R[i]$ is the maximum amount of commodity i that may be consumed. We would want $-y[i] \leq R[i]$ for all i. Indeed, if $y[i] \geq 0$, so that commodity i is produced, then we don't care about $R[i]$; in this case the inequality holds trivially. If $y[i] < 0$, then $-y[i]$ is the amount of commodity i used, and so we have to have $-y[i] \leq R[i]$. This inequality is $y + R \geq \mathbb{O}$. If there is a commodity that is used by each activity and not produced (e.g. labor), then this bounds each entry of x above, and so the domain of x is closed and bounded.

The set-up we have here is the basis for several optimization problems. For instance, suppose that we make chairs, using two activities (lumber work and finish work), and involving five commodities: labor, (raw) wood, milled (wood), fuel, and chairs. Here is the matrix $[A|R|C]$, where A, R are as above, and C will be explained momentarily.

	lumber	finish	available	cost factor
chairs	0	3	0	20
fuel	1	-1	1	3
milled	1	-2	0	1
wood	-2	-0.2	1	2
labor	-1	-3	5	3

Each commodity has a cost associated with it, in the form $c \cdot \sqrt{z}$, where c is the cost factor and z is the amount of the commodity. In our calculation of commodities coming from activities, if $y_i \geq 0$, then the cost gives rise to income $c_i \cdot \sqrt{y_i}$, where c_i is the cost factor. If commodity $y_i < 0$, then we pay the cost. In that case, let the cost be figured as *negative income*, so that our income is $-c_i \cdot \sqrt{-y_i}$. Our net income will be the sum of the income from each commodity.

It might help to show you how the two cases of the function of y_i can be represented by one formula in Excel:

$$=\texttt{IF}(y_i\texttt{<0},-c_i\texttt{*SQRT(}-y_i\texttt{)},c_i\texttt{*SQRT(}y_i\texttt{))}$$

Here, y_i and c_i stand for the cells that contain those numbers. This formula has an `if-then-else` structure.

◇ **Problem 91**
★ In the chair making example, find levels of the activities to maximize the sum of the commodity costs.
◇

CHAPTER 5

Convex Sets and Functions

1. Introduction

We are about to define what it means for a set to be convex and for a function to be convex. The two uses of the word *convex* are related but not in an obvious way; we will want to be sure we understand each of the two definitions on its own.

Suppose that $p, q \in \mathbb{R}^n$. The *standard segment* from p to q is the function

$$r(t) = (1 - t) \cdot p + t \cdot q \quad \text{for} \quad 0 \leq t \leq 1$$

We see that $r(0) = p$ and $r(1) = q$. The image of r is the line segment from p to q in $\mathbb{R}^n$. The function r can also be written like this.

$$r(t) = p + t \cdot (q - p)$$

Thus, it is an *affine function* – a linear function plus a constant.

A subset S of $\mathbb{R}^n$ is *convex* if whenever $p, q \in S$, all points on the standard segment from p to q are in S. In other words, if $p, q \in S$ and if $0 \leq t \leq 1$, then $(1 - t) \cdot p + t \cdot q$ is an element of S.

◇ **Problem 92**
A subset of the real numbers is convex if and only if it is an interval (open, closed, or half-open).
◇

◇ **Problem 93**
Let $x \in \mathbb{R}^n$ and $b > 0$. Show that the open disk $B(x, b)$ is convex, as is the closed disk $D(x, b)$.

117

Problem 94

Let V be a convex subset of $\mathbb{R}^n$, and let $b \in \mathbb{R}^n$. Then the set of $v + b$ for all $v \in V$ is convex.

$\Diamond$

 Problem 95

Let V be a convex subset of $\mathbb{R}^n$, and let $\alpha \in \mathbb{R}$. Define V' to be the set of $\alpha \cdot v$ for all $v \in V$. Show that V' is convex.

$\Diamond$

The feasible set of a linear program is convex.

 Problem 96

Let A be an $m \times n$ matrix and let B be $m \times 1$. Prove that the set of $n \times 1$ matrices X such that $AX \leq B$ is convex.

$\Diamond$

 Problem 97

Show that the intersection of convex sets is convex (or empty). Find two convex sets whose union is **not** convex.

$\Diamond$

The definition of the set W in the next problem is difficult. In working the problem, follow the definition of *convex* carefully. The problem says that the inverse image of a convex set under an affine map is convex.

 Problem 98

Let A be an $m \times n$ matrix, let B be an $m \times 1$ matrix, and let V be a convex subset of $\mathbb{R}^m$. Let W be the set of $x \in \mathbb{R}^n$ such that $Ax + B \in V$. Then W is a convex set.

$\Diamond$

Now we define what it means for a *function* to be convex. Suppose that V is a convex subset of $\mathbb{R}^n$ for some n, and let $f : V \to \mathbb{R}$. (Note that f is real-valued.) Then f is *convex* if for all $p, q \in V$, we have

(5.1) $f((1 - t) \cdot p + t \cdot q) \leq (1 - t) \cdot f(p) + t \cdot f(q)$ for $0 \leq t \leq 1$

When we want to emphasize the domain V of f, we say that f is *convex on V*.

When $f : \mathbb{R} \to \mathbb{R}$, convexity can be observed from the graph $y = f(x)$. In equation (5.1), the right side traces a *secant segment* from $f(p)$ to $f(q)$ as t goes from 0 to 1. The left side of (5.1) gives the y-values for f over the same set of x-values. Thus, (5.1) says that secant segments lie on or above the graph of the function.

The inequality (5.1) is an equality *for every function* when $t = 0$ or $t = 1$ or $p = q$. Thus, when we check whether a function is convex, it suffices to let $p \neq q$ and $0 < t < 1$.

Here is a general example: affine functions are convex.

◇ Problem 99
Let A be a $1 \times n$ matrix and let $b \in \mathbb{R}$. Then the function $A \cdot x + b$ is convex.
◇

◇ Problem 100
If V is a convex subset of $\mathbb{R}^n$, and if f, g are both convex on V, then so is $f + g$.
◇

◇ Problem 101
Show that all scalar multiples of a convex function are convex.
◇

◇ Problem 102
Let V be a convex subset of $\mathbb{R}^n$ and let $f : V \to \mathbb{R}$ be convex. Show that the function $f(-x)$ is convex.
◇

◇ **Problem 103**

Define $f : [0, 1] \to \mathbb{R}$ by $f(x) = x^2$ for $0 \leq x < 1$ and $f(1) = 2$. Show that f is convex. (Hint: the graph makes this pretty clear. Note that f is not continuous at $x = 1$!)

◇

The composite of an affine function with a convex function is convex.

PROPOSITION 5.1. *Let A be an $m \times n$ matrix, let B be an $m \times 1$ matrix. Let V be a convex subset of $\mathbb{R}^m$, and let W be the set of $x \in \mathbb{R}^n$ such that $A \cdot x + B \in V$. Finally suppose that $f : V \to \mathbb{R}$ is convex. Then the function $f(Ax + B)$, mapping W into $\mathbb{R}$ is convex.*

PROOF. Note that a problem above showed that W is convex.

Let $p, q \in W$, let $t \in [0, 1]$, and define $s = (1 - t) \cdot p + t \cdot q$. We need to show that

$$f(As + B) \leq (1 - t) \cdot f(Ap + B) + t \cdot f(Aq + B)$$

We compute

$$
\begin{aligned}
As + B &= A \cdot \big[(1 - t)p + tq\big] + B \\
&= (1 - t) \cdot A \cdot p + t \cdot A \cdot q + B \\
&= (1 - t) \cdot A \cdot p + t \cdot A \cdot q + (1 - t) \cdot B + t \cdot B \\
&= (1 - t) \cdot \big[Ap + B\big] + t \cdot \big[Aq + B\big]
\end{aligned}
$$

and so, using that f is convex,

$$
\begin{aligned}
f(As + B) &= f\Big((1 - t) \cdot \big[Ap + B\big] + t \cdot \big[Aq + B\big]\Big) \\
&\leq (1 - t) \cdot f(Ap + B) + t \cdot f(Aq + B)
\end{aligned}
$$

as needed.   ∎

Here is one way to relate convex sets and functions.

◇ **Problem 104**

Let $V \subseteq \mathbb{R}^n$ be a convex set, and let $f : V \to \mathbb{R}$. Define S to be the set of all $(x, y) \in \mathbb{R}^{n+1}$ such that $x \in V$ and $y \geq f(x)$. Show that S is a convex set if and only if f is a convex function.

◇

We will now prove that a function is convex if and only if it is convex along each standard segment, as a function of one variable on that segment. The proof involves a real-valued function $f(r(t))$; we noted above that such a function is convex if and only if its secant segments lie on or above its graph.

PROPOSITION 5.2. *Let V be a convex subset of $\mathbb{R}^n$, and let $f : V \to \mathbb{R}$. Then f is convex if and only if $f(r(t))$ is convex for every standard segment $r(t)$ in V.*

PROOF. Let $p, q \in V$, and let $r(t)$ be the standard segment from p to q. The function $r(t)$ is affine and if f is convex, then $f(r(t))$ is convex by Proposition 5.1.

Conversely, if $f(r(t))$ is convex, then if $0 \leq t \leq 1$, we have

$$f((1-t)p + tq) = f(r(t)) = f(r((1-t) \cdot 0 + t \cdot 1))$$
$$\leq (1-t) \cdot f(r(0)) + t \cdot f(r(1))$$
$$= (1-t) \cdot f(p) + t \cdot f(q)$$

and this is exactly that f is convex, since r is arbitrary.   ∎

The graphing property involving secant segments makes it easy to see that a function of one variable is convex if and only if it is concave up. A function is concave up if its first derivative is increasing – if its second derivative is non-negative. The proof of this apparent fact takes a little time.

PROPOSITION 5.3. *Let J be an interval in the reals, and let $f : J \to \mathbb{R}$. If f is differentiable, then it is convex if and only if its derivative is increasing. If f'' exists, then f is convex if and only if $f''(x) \geq 0$ for all $x \in J$.*

PROOF. Let $p, q \in J$ and we might as well assume that $p < q$. Let t be in the open interval $(0, 1)$. Define $s = (1 - t) \cdot p + t \cdot q$. We examine the difference

$$(1 - t) \cdot f(p) + t \cdot f(q) - f(s) = (1 - t) \cdot \left(f(p) - f(s) \right) + t \cdot \left(f(q) - f(s) \right)$$

We apply the Mean Value Theorem to the two differences that result. There is a number a such that $p < a < s$ and $f(p) - f(s) = f'(a) \cdot (p - s)$, and there is a number b such that $s < b < q$ and $f(q) - f(s) = f'(b) \cdot (q - s)$. Notice that $p < a < s < b < q$ shows that $a < b$. We have

$$(1 - t) \cdot f(p) + t \cdot f(q) - f(s) = (1 - t) \cdot f'(a) \cdot (p - s) + t \cdot f'(b) \cdot (q - s)$$

The definition of s shows that

$$p - s = -t \cdot (q - p) \quad \text{and} \quad q - s = (1 - t) \cdot (q - p)$$

Thus,

$$(1 - t) \cdot f(p) + t \cdot f(q) - f(s)$$
$$= (1 - t) \cdot f'(a) \cdot (-t) \cdot (q - p) + t \cdot f'(b) \cdot (1 - t) \cdot (q - p)$$
$$= (1 - t) \cdot t \cdot (q - p) \cdot \left(f'(b) - f'(a) \right)$$

We remarked that $a < b$. If $f'(x)$ is increasing, then $f'(a) \leq f'(b)$, so that $(1 - t) \cdot f(p) + t \cdot f(q) - f(s) \geq 0$. This shows that f is convex.

The converse case takes some fussing.[1] Suppose that f is convex. If f' is not increasing, and we can find $a < d$ on J such that $f'(a) > f'(d)$. Using the limit that defines the derivative, we can find $a < b < c < d$ with $b - a = d - c$, and

$$\frac{f(b) - f(a)}{b - a} > \frac{f(d) - f(c)}{d - c}$$

Clearing the equal denominators, we see that

(5.2) $$f(b) + f(c) > f(a) + f(d)$$

Since $a < b < d$, there is $0 < t < 1$ such that $b = (1 - t) \cdot a + t \cdot d$. Since $b - a = d - c$, we have

---

[1]If f is convex, then we have $f'(a) \leq f'(b)$ for the a, b that arise as in the previous argument. The trouble is to show that a, b are *arbitrary*.

$$c = d + a - b = t \cdot a + (1 - t) \cdot d$$

Since f is convex, we have

$$f(b) \leq (1 - t) \cdot f(a) + t \cdot f(d)$$
$$f(c) \leq t \cdot f(a) + (1 - t) \cdot f(d)$$

Adding these inequalities, we contradict (5.2). This proves that f' is increasing, after all.

To deduce from the second derivative, if $f''(x) \geq 0$, then f' is increasing and f is convex. If f is convex, then f' is increasing, and $f'' \geq 0$.   ∎

 Problem 105
Show that x^2 and $(x - 1)^2$ are convex on $[0, 1]$, but their product is *not convex* on $[0, 1]$.
◇

 Problem 106
Suppose $f(x)$ is a decreasing function on the reals that satisfies the *law of diminishing returns*: increases in the independent variable lead to smaller and smaller decreases of $f(x)$. Show that $f(x)$ is convex.
◇

◇ **Problem 107**
Here is another version of diminishing returns! Suppose that $f(x)$ is an increasing function and that increases in the independent variable lead to smaller and smaller increases in $f(x)$. Show that $-f(x)$ is convex.
◇

For the typical *utility function* $f(x)$ of economics, we usually assume that $-f(x)$ is convex. Such a function is called *concave*.

2. The Hessian

If $f(x)$ is a function of *more than one variable*, is there a result similar to Proposition 5.3 that allows us to tell whether f is convex? The answer is "Yes," with the *Hessian*, introduced momentarily, playing the role of the second derivative.

A function is C^2 if all its second partial derivatives are defined and continuous. When f is C^2, we define the *Hessian* $\mathbb{H}$ of f as the $n \times n$ matrix of functions whose entries are these partial derivatives:

$$\mathbb{H}[i, j] = \frac{\partial^2 f}{\partial x_i \partial x_j}$$

It is a fact that since the partial derivatives are continuous, the order of taking the two partial derivatives is immaterial:[2]

$$\frac{\partial^2 f}{\partial x_i \partial x_j} = \frac{\partial^2 f}{\partial x_j \partial x_i}$$

and therefore $\mathbb{H}[i, j] = \mathbb{H}[j, i]$ for all i, j, so that we have $\mathbb{H}^T = \mathbb{H}$. The domain of the Hessian is the same as the domain of f.

 Problem 108

Find the Hessian of the function $x^2 \cdot y + x \cdot y^3$.

$\Diamond$

 Problem 109

Prove that $f : \mathbb{R}^n \to \mathbb{R}$ has constant Hessian $2 \cdot A$ if and only if there is an $n \times n$ matrix A, with $A = A^T$, a $1 \times n$ matrix B, and a number c such that

$$f(X) = X^T \cdot A \cdot X + B \cdot X + c \quad \text{for all} \quad X \in \mathbb{R}^n$$

(Hint: a problem on p.82 gives a suggested expansion of $X^T A X$.) Such a function f is called a *quadratic function*.

$\Diamond$

---

[2]You can find a proof in [**5**, p.208]

The proof of the following is a direct calculation; it shows how the second derivative is computed using the Hessian. We write $\mathbb{H}(x)$ for the numerical matrix obtained by evaluating the entries of $\mathbb{H}$ at x.

PROPOSITION 5.4. *Let $V \subseteq \mathbb{R}^n$ be a convex set, and let $f : V \to \mathbb{R}$ be C^2. Let $p, q \in V$, and let $r(t)$ be the standard segment from p to q. Then $f(r(t))'' = (q - p)^T \cdot \mathbb{H}(r(t)) \cdot (q - p)$. In particular, $f(r(t))''$ exists and is continuous.*

PROOF. Let $v = q - p$, and the definition of $r(t)$ shows that $Dr = v$. The Chain Rule computes

$$D_t\Big[f(r(t))\Big] = Df(r(t)) \cdot Dr = Df(r(t)) \cdot v$$

Note that Df is $1 \times n$ and v is $n \times 1$, so that their product is a real number.

We write out $Df(r(t)) \cdot v$ in terms of the partial derivatives of f with respect to its variables x_j, with $1 \le j \le n$.

$$Df(r(t)) \cdot v = \sum_{j=1}^{n} \frac{\partial f}{\partial x_j}(r(t)) \cdot v_j$$

To compute the second derivative with respect to t, we apply the Chain Rule to each term in this sum, using the same notation for the variables:

$$D_t\left[\frac{\partial f}{\partial x_j}(r(t)) \cdot v_j\right] = \sum_{i=1}^{n} \frac{\partial^2 f}{\partial x_i \partial x_j} \cdot v_i \cdot v_j$$

(We drop the $r(t)$; all the second partial derivatives are evaluated at $r(t)$.)

Thus,

$$\Big[f(r(t))\Big]'' = D_t\left[\sum_{j=1}^{n}\frac{\partial f}{\partial x_j} \cdot v_j\right] = \sum_{j=1}^{n}\sum_{i=1}^{n} \frac{\partial^2 f}{\partial x_i \partial x_j} v_i \cdot v_j$$

$$= \sum_{j=1}^{n}\sum_{i=1}^{n} \mathbb{H}[i, j] \cdot v_i \cdot v_j = \sum_{j=1}^{n} \left(v^T \cdot \mathbb{H}\right)[1, j] \cdot v_j = v^T \cdot \mathbb{H} \cdot v$$

as needed. Furthermore, since the entries in $\mathbb{H}$ are continuous, we see that $f(r(t))''$ is continuous. ∎

The next result shows how the Hessian can be used to tell whether a function of several variables is convex. Note that the domain in the following is an open set: we need to be able to move around in all directions from a given point.

PROPOSITION 5.5. *Let $V \subseteq \mathbb{R}^n$ be an open convex set, and let $f : V \to \mathbb{R}$ be C^2. Let $\mathbb{H}$ be the Hessian for f. Then f is convex if and only if $v^T \cdot \mathbb{H}(p) \cdot v \geq 0$ for all $p \in V$ and all $n \times 1$ matrices v.*

PROOF. By Proposition 5.2, f is convex if and only $f(r(t))$ is convex for each such $r(t)$. By Proposition 5.3, $f(r(t))$ is convex if and only if its second derivative is non-negative. Proposition 5.4 computes the second derivative as $(q-p)^T \cdot \mathbb{H}(r(t)) \cdot (q-p)$, where $r(t)$ goes from p to q. Conclusion: f is convex if and only if

(5.3) $$(q-p)^T \cdot \mathbb{H}(r(t)) \cdot (q-p) \geq 0$$

for all $p, q \in V$, where $r(t)$ is the standard segment from p to q.

Suppose that $v^T \cdot \mathbb{H}(p) \cdot v \geq 0$ for all $p \in V$ and $v \in \mathbb{R}^n$. Then clearly (5.3) holds, and so f is convex.

Now suppose that f is convex, and let $p \in V$ and $v \in \mathbb{R}^n$. Because V is open,[3] there is a positive number δ such that $p + \delta \cdot v \in V$. Let $q = p + \delta \cdot v$, and let $r(t)$ be the standard segment from p to q. We know that $(q-p)^T \cdot \mathbb{H}(r(t)) \cdot (q-p) \geq 0$ for all $t \in [0,1]$. Taking $t = 0$ and remembering v, we have

$$\delta \cdot v^T \cdot \mathbb{H}(p) \cdot \delta \cdot v = \delta^2 \cdot v^T \cdot \mathbb{H}(p) \cdot v$$

Because of (5.3), this expression is non-negative, and so $v^T \cdot \mathbb{H}(p) \cdot v \geq 0$, as needed. ∎

Proposition 5.5 calls attention to a property that a matrix A might have: if $A = A^T$ and $x^T \cdot A \cdot x \geq 0$ for all $x \in \mathbb{R}^n$, we say that A is *positive semi-definite*. Proposition 5.5 says that f is convex if and only if its Hessian is positive semi-definite.

---

[3]That v is arbitrary is what we meant by saying we can move around in all directions.

Problem 110

Suppose that A is 2×2 and $A = A^T$, and prove that A is positive semi-definite if and only if $A[1,1] \geq 0$ and $A[2,2] \geq 0$ and $A[1,1] \cdot A[2,2] \geq A[1,2]^2$. (Hint: let $f(X) = X^T \cdot A \cdot X$ for all $X \in \mathbb{R}^2$. Consider separately the cases $A[1,1] = 0$ and $A[1,1] \neq 0$, and play around with the coordinates in X.

$\diamondsuit$

 Problem 111

Let $f : \mathbb{R}^2 \to \mathbb{R}$ by $f(x,y) = x^2 + x \cdot y + y^2 + x$. Prove that f is convex.

$\diamondsuit$

 Problem 112

For each of the two matrices A given here, define $f(X) = X^T A X$, and figure out whether the function is convex.

$$\begin{pmatrix} 1 & 3 \\ 3 & 1 \end{pmatrix} \quad \text{and} \quad \begin{pmatrix} 2 & 0 & 0 \\ 0 & 1 & 0 \\ 0 & 0 & 3 \end{pmatrix}$$

$\diamondsuit$

$\diamondsuit$ **Problem 113**

Define V to be the set of points $(x,0) \in \mathbb{R}^2$. Let $f(x,y) = x^2 - y^2$ define f on V. Show that V is convex and $f : V \to \mathbb{R}$ is convex. Let $\mathbb{H}$ be the Hessian of f. Find $v \in \mathbb{R}^2$ such that $v^T \cdot \mathbb{H} \cdot v < 0$. (This shows that Proposition 5.5 is false if the hypothesis "V is open" is dropped.)

$\diamondsuit$

There are various theoretical tests to determine whether a given $n \times n$ matrix is positive semi-definite. We mention three of them. Each can be used in various contexts, depending on the specific matrix.

The Factor Test.   The matrix A is positive semi-definite if and only if there is a matrix C such that $C^T \cdot C = A$. (Notice that C does not have to be square.)

◇ **Problem 114**

Let C be a matrix. Show that $C^T \cdot C$ is positive semi-definite. (Hint: $X^T \cdot C^T \cdot C \cdot X$ is the square of a norm.)

◇

   Our next two tests involve terms we will not define: the *determinant* of a matrix and the *eigenvalues* of a matrix. Many numerical programs will approximate the determinant and/or the eigenvalues[4] of a given matrix.

The Eigenvalue Test.   Let A be $n \times n$ with $A = A^T$. Then the eigenvalue of A are real numbers and A *is positive semi-definite if and only if the eigenvalues are all non-negative.*

◇ **Problem 115**

Let A be an $n \times n$ matrix with $A = A^T$. Prove that if A has a negative number eigenvalue, then it is not positive semi-definite.

◇

   The next test involves two statements; we point out that these statements are not converses.

Determinant Test for positive semi-definite.   For each integer i with $1 \leq i \leq n$, define A_i to be the $i \times i$ matrix consisting of the upper left-hand corner of A (so that $A_i[j, k] = A[j, k]$ for all $1 \leq j, k \leq i$). *If each A_i has positive determinant, then A is positive semi-definite.* Also, *if A is positive semi-definite, then each A_i has a non-negative determinant.*

---

[4]The complex number α is an eigenvalue of the $n \times n$ matrix A if $A \cdot v = \alpha \cdot v$ for some non-zero $n \times 1$ matrix v with complex entries.

◇ **Problem 116**

Let $A = \begin{pmatrix} 0 & 0 \\ 0 & -1 \end{pmatrix}$. Notice that A_1 and A_2 have *non-negative* determinants, but A is not positive semi-definite.

◇

◇ **Problem 117**

Let A be 2×2 with $A = A^T$. In the notation of the determinant test, suppose that A_1 and A_2 have positive determinants. Show that the eigenvalues of A are positive.

◇

CHAPTER 6

Convex Optimization

1. Interior Extremes

In Chapter 5, we defined convex sets and convex functions. Let V be an open convex subset of $\mathbb{R}^n$, and let $f : V \to \mathbb{R}$ be a convex function. Let m be a positive integer, and let $g : V \to \mathbb{R}^m$ be such that each coordinate function g_i is convex. Then the following problem is a *convex program*: minimize $f(x)$ such that $g(x) \le \mathbb{O}$. Note that every linear program in primal form is a convex program! Thus, the application of convex programs include all the linear programs – and a good deal more besides. In some sense, convex programs are easier than arbitrary non-linear programs.

We begin with a converse to the Interior Extreme Theorem – this tells us how to solve a convex program when there are no constraints.

PROPOSITION 6.1. *Let V be an open, convex subset of $\mathbb{R}^n$, and suppose that $f : V \to \mathbb{R}$ is a C^1 convex function. If $p \in V$ and $Df(p) = \mathbb{O}$, then $f(p)$ is the minimum for f on V.*

PROOF. Let $q \in V$, and let $r(t)$ be the standard segment from p to q. Proposition 5.3 of Chapter 5 shows that the derivative of $f(r(t))$ is increasing. This derivative is $Df(r(t)) \cdot (q - p)$ and it has value $Df(p) = 0$ when $t = 0$. Thus, the derivative is non-negative for all t, and so $f(r(t))$ is increasing. It follows that $f(q) \ge f(p)$. ∎

Let A be $n \times n$ with $A = A^T$, and suppose that A is positive semi-definite. Define $f : \mathbb{R}^n \to \mathbb{R}$ by $f(x) = x^T \cdot A \cdot x$. The definition of positive semi-definite is exactly that f has minimum 0; of course $f(\mathbb{O}) = 0$. We have computed Df, and you should see that $Df(\mathbb{O}) = \mathbb{O}$.

131

◇ Problem 118

Let A be $n \times n$ with $A = A^T$, and suppose that A is positive semi-definite. Let B be $1 \times n$. Define $f : \mathbb{R}^n \to \mathbb{R}$ by $f(x) = x^T \cdot A \cdot x + B \cdot x$. Show that f has a minimum if and only if there is an $n \times 1$ matrix v such that $2 \cdot v^T \cdot A = -B$. If v exists, then $f(v)$ is a minimum for f.
◇

◇ Problem 119

Find the minimum of $x^T \cdot A \cdot x + B \cdot x$ where

$$A = \begin{pmatrix} 2 & 1 \\ 1 & 3 \end{pmatrix} \quad \text{and} \quad B = \begin{pmatrix} -1 & 2 \end{pmatrix}$$

◇

◇ Problem 120

Find an $n \times n$ positive semi-definite matrix A with $A = A^T$, and a $1 \times n$ matrix B such that $f(x) = x^T \cdot A \cdot x + B \cdot x$ has no minimum.
◇

2. Kuhn-Tucker for Convex Programs

In Chapter 4 we *assumed* we had an extreme point for a function under Hypothesis 4.1 and *proved* that the Kuhn-Tucker Conditions hold. In convex programs, we do not need the technical rank condition to get Kuhn-Tucker, although we do need to know that the constraints allow strict inequalities. There is a version of the following result for non-differentiable functions; we will not pursue this more technical direction.

PROPOSITION 6.2. *Let V be a convex open subset of $\mathbb{R}^n$ and let $f : V \to \mathbb{R}$ be convex and C^1 and let $g : V \to \mathbb{R}^m$ have convex, C^1 coordinates, and suppose there is $q \in V$ with $g(q) < \mathbb{O}$. Suppose that $p \in V$ is a solution to the problem: minimize $f(x)$ such that $g(x) \leq \mathbb{O}$. Then the Kuhn-Tucker conditions hold at p.*

PROOF. Assume that $g(p) = \mathbb{O}$. Let $u = q - p$.

Claim 1. We have $Dg(p) \cdot u < \mathbb{O}$.

Let $1 \leq i \leq m$. Since $g_i(p) = 0 > g_i(q)$, the convex function $h(t) = g_i(p + t \cdot u)$ has negative derivative for some t with $0 < t < 1$. Since $h'(t)$ is increasing, we see that $h'(0) < 0$; that's $Dg_i(p) \cdot u < 0$. ∎

Claim 2. If $v \in \mathbb{R}^n$ and $Dg(p) \cdot v \leq \mathbb{O}$, then $Df(p) \cdot v \geq 0$.

Suppose that $Dg(p) \cdot v \leq \mathbb{O}$ and $Df(p) \cdot v < 0$. For each positive number α, Claim 1 shows that

$$Dg(p) \cdot (v + \alpha \cdot u) \leq Dg(p) \cdot \alpha \cdot u < 0$$

Since $Df(p) \cdot v < 0$, we can choose α small enough so that

$$Df(p) \cdot (v + \alpha \cdot u) < 0$$

Let $w = v + \alpha \cdot u$, and we will consider points $y = p + \beta \cdot w$ for $\beta > 0$.

Since V is open, we can find a make β small enough so that $y \in V$. For each i with $1 \leq i \leq m$, we have $Dg_i(p) \cdot w < 0$, and so we can make β smaller so that

$$g_i(y) < g_i(p) = 0$$

In other words, y is feasible for the convex program. Since $Df(p) \cdot w < 0$, we can decrease β again, so that $f(y) < f(p)$, and that contradicts the minimality of p. The Claim holds. ∎

Claim 2, allows us to apply Proposition 2.8 to $Df(p)$ and the columns of $-Dg(p)$. Reversing the sign on $Dg(p)$, that result finds a $1 \times m$ matrix $\lambda \leq \mathbb{O}$ such that $Df(p) = \lambda \cdot Dg(p)$. We have established the Kuhn-Tucker conditions in the case that $g(p) = \mathbb{O}$.

We handle the general case exactly as in the general, non-linear situation. If $g = (G, H)$, where G is the zero part of g with respect to p, and H is the non-zero part, we get λ as above from G, and we let the multipliers corresponding to H be 0. ∎

Next we establish the converse: if the Kuhn-Tucker Conditions hold, then we have a minimum.

THEOREM 6.3. *Let V be an open convex subset of $\mathbb{R}^n$, let $f : V \to \mathbb{R}$ be C^1 and convex, and let $g : V \to \mathbb{R}^m$ be C^1 with each g_i convex. Suppose that the Kuhn-Tucker conditions hold at $p \in V$. Then $f(p)$ is a minimum for f on the feasible set of the convex program defined by f, g.*

PROOF. Let λ be the Lagrange Multiplier, a $1 \times m$ matrix with non-positive entries. The function $f - \lambda \cdot g$ is defined and real-valued on the convex set V. We claim that $f - \lambda \cdot g$ is a convex function. Indeed, by a problem in Chapter 5, non-negative multiples of convex functions are convex, and sums of convex functions are convex. Since $\lambda[j] \leq 0$ for each j, we see that $f - \lambda \cdot g$ is convex.

Next, use the fact that $Df(p) = \lambda \cdot Dg(p)$ to see that $D(f - \lambda \cdot g)(p) = 0$. By Proposition 6.1, $f(p) - \lambda \cdot g(p)$ is the minimum of $f - \lambda \cdot g$ on V. Complementary Slackness shows that $\lambda \cdot g(p) = 0$, and so the value of $f - \lambda \cdot g$ at p is $f(p)$.

Let $v \in V$ be feasible, so that $g(v) \leq \mathbb{O}$, and since $-\lambda \geq \mathbb{O}$, we see that $-\lambda \cdot g(v) \leq 0$. It follows that $f(v) - \lambda \cdot g(v) \leq f(v)$. Now we have

$$f(p) = f(p) - \lambda \cdot g(p) \leq f(v) - \lambda \cdot g(v) \leq f(v)$$

This proves that $f(p) \leq f(v)$, and we see that $f(p)$ is a minimum for f on the feasible set. ∎

◇ Problem 121

Let C be $1 \times n$, and let E be an $n \times n$ invertible matrix. Let r be a number. Show that this problem is convex: minimize $C \cdot X$ such that $X^T \cdot E^T \cdot E \cdot X = r$. Find the solution.

◇

◇ Problem 122

Suppose that the output Q of an economy satisfies a *Cobb-Douglas function* $Q = L^a \cdot C^b$ where L is labor, C is capital, and a, b are positive constants with $a + b \leq 1$. (The domain of Q is the set of all (L, C) where $L \geq 0$ and $C \geq 0$.)
(a) Show that $-Q$ is a convex function[1] of L, C.

---

[1] We have mentioned that the negative of a convex function is called a *concave* function.

(b) Maximize Q such that $L + C = d$ where d is a positive constant.

◇

You might recall Proposition 2.6, which says that the Kuhn-Tucker conditions are necessary and sufficient for a solution to a linear program. Therefore, we can describe the simplex algorithm abstractly by saying that it finds a solution to Kuhn-Tucker. For a convex program, solving the Kuhn-Tucker conditions also solves the problem, although we do not have a definitive algorithm.

3. Applications

3.1. Consumption Theory. We have a consumer with w dollars available to buy a combination of n goods. There is a $1 \times n$ price vector p, so that $p[i]$ is the unit price of good i. If x is the $n \times 1$ vector of purchases, then our consumer can't spend more than w, and so $p \cdot x \leq w$. The consumer has a *utility function* $u(x)$ that quantifies the overall benefit coming from the purchases indicated in x. We assume that $-u(x)$ is convex (that $u(x)$ is concave). The consumption problem: maximize $u(x)$ such that $p \cdot x \leq w$ and $x \geq \mathbb{0}$.

The maximum of u occurs with the minimum of $-u$, and we see that the consumption problem is a convex program.

◇ **Problem 123**
Let V be the set of $x \in \mathbb{R}^n$ such that $x > \mathbb{0}$. Let $\alpha_1, \ldots, \alpha_n$ be positive numbers, and define $u : V \to \mathbb{R}$ by

$$u(x) = \sum_{j=1}^{n} \alpha_j \cdot \ln(x[j])$$

(a) Show that $-u$ is convex.
(b) Solve the consumption problem for this utility function.
(c) Find the specific solution in case

$$(\alpha_1, \alpha_2, \alpha_3) = \begin{pmatrix} 0.4 & 2 & 0.6 \end{pmatrix} \quad \text{and} \quad p = \begin{pmatrix} 2 & 5 & 3 \end{pmatrix} \quad \text{and} \quad w = 300$$

3.2. Quadratic Programs. A *quadratic program* has this form: minimize $X^T \cdot A \cdot X + B \cdot X$ such that $C \cdot X \le E$, where the variables are X, and the matrix A satisfies $A = A^T$ and is positive semi-definite. Bookkeeping: if X is $n \times 1$ and C is $m \times n$, then A is $n \times n$ and B is $1 \times n$ and E is $m \times 1$.

We note that constraints in equation form can be converted to the primal form required by the usual expedient of writing each equation as two inequalities. If the variables are required to be non-negative, then those inequalities are included in $C \cdot X \le E$.

By a problem in Chapter 5, the Hessian of the objective is $2 \cdot A$ and since this matrix is positive semi-definite, the objective is convex. The feasible vectors form a convex set as well, and so, Theorem 6.3 is in force: if the Kuhn-Tucker conditions hold, then we have a solution.

$\diamond$ Problem 124

Show that the Kuhn-Tucker conditions for the quadratic program are these: There is an $n \times 1$ matrix V and a $1 \times m$ matrix $L \le \mathbb{O}$ such that $C \cdot V \le E$ and $2 \cdot V^T \cdot A + B = L \cdot C$ and $L \cdot C \cdot V = L \cdot E$.
$\diamond$

We can create an $m \times 1$ matrix S of non-negative slack variables, and write $C \cdot X \le E$ as $C \cdot X + S = E$ as usual. Let $Y = -L$. Now consider the LP: minimize $S[1]$ such that $C \cdot X + S = E$ and $2 \cdot X^T \cdot A + B = -Y \cdot C$ and $Y \ge \mathbb{O}$ and $S \ge \mathbb{O}$.

The LP captures the Kuhn-Tucker conditions except for the equation $Y \cdot C \cdot X = Y \cdot E$. Notice that this equation is $Y \cdot S = 0$. P. Wolfe[2] proved that the simplex algorithm can be modified to solve the LP in such a way as to get this extra condition as well. In Wolfe's modification, we make sure that $Y[j]$ and $S[j]$ are not both basic at every step of the algorithm. (If one of them is basic at some point, we do not look at replacements that would make the other one basic.) Wolfe proved that the algorithm still finds a solution under this modification. We will not prove Wolfe's theorem.

Here are some sample applications of quadratic programming.

---

[2]See the paper [**7**].

Investment Correlations.

We have three possible investments A, B, C in which we will invest a total of \$100. Table A below gives the expected rate of return on each investment and the variance of that expected rate. Table B gives the covariance between the various rates. Decide how to invest \$100 so as to minimize the variance of the return while attaining an expected return of at least \$9.50.

Table A	A	B	C
expected rate	.07	.09	.10
variance	.01	.008	.009

Table B

A,B	A,C	B,C
0.007	0.003	0.001

Let the amount invested in A, B, C be X_1, X_2, X_3, respectively. The expected return E on the three investments is the sum of the product of the amount invested times the expected return for that investment. Denoting the rates by r_1, r_2, r_3, we have

$$E = \sum_{j=1}^{3} r_j \cdot X_j$$

The expected return is the ratio of E to the amount invested. To compute the variance V of the return, we let v_j be the variance of the j-th investment, and we let $c_{i,j}$ be the covariance between investments i and j (where $i < j$). Then

$$V = \sum_{j=1}^{3} v_j \cdot X_j^2 + 2 \cdot \sum_{1 \le i < j \le 3} c_{i,j} \cdot X_i \cdot X_j$$

Define the 3×3 matrix A with $A[i, i] = v_i$ for $1 \le i \le 3$ and $A[i, j] = A[j, i] = c_{i,j}$ when $1 \le i < j \le 3$. The variance of the return is necessarily non-negative, and so A is positive semi-definite. It follows that the investment problem is a quadratic program.

$\Diamond$ **Problem 125**

★ Solve the investment problem.

Curve Fitting with Constraints.

In Chapter 4 we described curve-fitting, where matrices A, B are given and we find X to minimize $|A \cdot X - B|^2$. Here we consider additional constraints on X: suppose we need $C \cdot X \leq E$ for some $m \times n$ matrix C and $m \times 1$ matrix E.

◇ **Problem 126**

Show that the curve-fitting problem with the constraint $C \cdot X \leq E$ is a quadratic program. (Hint: write $|A \cdot X - B|^2$ as $(AX - B)^T \cdot (AX - B)$. Use the Factor Test for positive semi-definiteness.)

◇

◇ **Problem 127**

★ Find the plane $z = a \cdot x + b \cdot y$ in $\mathbb{R}^3$ (where (x, y, z) is a point on the plane) such that $a \geq b$ and of best fit to the following data points: $(1, 2, 3)$, $(4, 5, 6)$, $(-1, 0, 3)$, $(-2, 1, -1)$.

◇

4. Local Minima

Proposition 2.6 and Theorem 6.3 give results that say, "If the Kuhn-Tucker conditions hold, then we have a minimum." We wish to prove such a result in general, but we will have a *local minimum*, a minimum when the objective is restricted to an open disk. In the open disk we will consider, the Hessian of the objective will be *positive definite*, a condition stronger than being positive semi-definite.

Here are the details. An $n \times n$ matrix A with $A = A^T$ is *positive definite* if $v^T \cdot A \cdot v > 0$ for all $v \in \mathbb{R}^n$ with $v \neq \mathbb{O}$. If A is positive definite, it is also positive semi-definite, but not necessarily vice versa. The matrix

$$\begin{pmatrix} 0 & 0 \\ 0 & 1 \end{pmatrix}$$

is positive semi-definite but not positive definite. Here is the fact we need about positive definite matrices.

◇ **Problem 128**
Suppose that the matrix $A = A^T$ is positive definite. Show that A has an inverse and that A^{-1} is positive definite, as well.

◇

PROPOSITION 6.4. *Let $V \subseteq \mathbb{R}^n$ be open, and let $f : V \to \mathbb{R}$ be C^2. Let $\mathbb{H}(x)$ be the Hessian of f. Let $p \in V$ and suppose that $\mathbb{H}(p)$ is positive definite. Then there is an open convex set $W \subseteq V$ with $p \in W$ and such that if $q \in W$, then $\mathbb{H}(q)$ is positive definite. In particular, f is convex on W.*

PROOF. Let S be the set of $y \in \mathbb{R}^n$ such that $|y| = 1$. For $y \in S$, we need to estimate

$$\sum_{i=1}^{n} |y[i]|$$

To do this, let $J \in \mathbb{R}^n$ have $J[i] = 1$ for all i, and our sum is the dot product $J \circ y$. Thus, the Cauchy-Schwarz inequality says

(6.1) $$\sum_{i=1}^{n} |y[i]| \leq |J| \cdot |y| = \sqrt{n}$$

For $y \in S$, define $h(y) = y^T \cdot \mathbb{H}(p) \cdot y$, and then $h : S \to \mathbb{R}$ is continuous. Since S is closed and bounded, the Extreme Value Theorem gives h a minimum δ on S. Since $\mathbb{H}(p)$ is positive definite, $\delta > 0$.

Since $\mathbb{H}$ is continuous, there is an open disk W centered at p such that if $x \in W$, then each entry of $\mathbb{H}(x)$ is within $\delta/(2n)$ of the corresponding entry of $\mathbb{H}(p)$. (We will see the reason for $\delta/(2n)$ shortly!) Notice that W is convex, being a disk.

Let $x \in W$, and we will show that $\mathbb{H}(x)$ is positive definite. Temporarily define $A = \mathbb{H}(x) - \mathbb{H}(y)$, so that $|A[i,j]| \leq \delta/(2n)$ for all i, j. Let $y \in S$ and use (6.1) to estimate

$$\left| y^T \cdot A \cdot y \right| = \left| \sum_{i,j} y[i] \cdot A[i,j] \cdot y[j] \right|$$

$$\leq \sum_{i,j} |y[i]| \cdot \frac{\delta}{2n} \cdot |y[j]|$$

$$= \left(\sum_{i=1}^{n} |y[i]| \right)^2 \cdot \frac{\delta}{2n} \leq n \cdot \frac{\delta}{2n} = \frac{\delta}{2}$$

In light of this, we estimate

$$y^T \cdot \mathbb{H}(x) \cdot y = y^T \cdot \mathbb{H}(p) \cdot y + y^T \cdot A \cdot y$$

$$\geq y^T \cdot \mathbb{H}(p) \cdot y - \left| y^T \cdot A \cdot y \right|$$

$$\geq \delta - \frac{\delta}{2} = \frac{\delta}{2} > 0$$

Now let $v \in \mathbb{R}^n$ with $v \neq \mathbb{O}$, and then $v/|v| \in S$, and so

$$v^T \cdot \mathbb{H}(x) \cdot v = |v|^2 \cdot (v/|v|)^T \cdot \mathbb{H}(x) \cdot (v/|v|) \geq |v|^2 \cdot \delta/2 > 0$$

Because the Hessian of the function f is positive definite on W, it is positive semi-definite on that set, and Proposition 5.5 says that f is convex on W. ∎

If $f(x,y) = x^3 + y^2$, then the Hessian of f at $(0,0)$ is positive semi-definite, but there is no open set containing $(0,0)$ in which the Hessian of f is positive semi-definite. Thus, *positive semi-definite* cannot be substituted for *positive definite* in Proposition 6.4.

Armed with the previous proposition, we can prove that the Kuhn-Tucker conditions, along with a positive definite Hessian, imply that a point is a local minimum for an optimization problem in primal form.

THEOREM 6.5. *Let $V \subseteq \mathbb{R}^n$ be open, and let $f : V \to \mathbb{R}$ be C^2. Let $g : V \to \mathbb{R}^m$ be C^2. Suppose that $p \in V$ and that $\lambda \in \mathbb{R}^m$ satisfies $Df(p) = \lambda \cdot Dg(p)$ and $\lambda \leq \mathbb{O}$ and $\lambda \cdot g(p) = 0$. Suppose also that the Hessian of $f(x) - \lambda \cdot g(x)$ is positive definite at p. Then there is an open set $W \subseteq V$ with $p \in W$ such that $f(p)$ is the minimum value of f over the set of $x \in W$ such that $g(x) \leq \mathbb{O}$.*

PROOF. By Proposition 6.4, there is a convex open set W with $p \in W$ and such that $f(x) - \lambda \cdot g(x)$ is convex on W. The equation $Df(p) = \lambda \cdot Dg(p)$ tells us that the derivative of $f(x) - \lambda \cdot g(x)$ is 0 at p, and so Proposition 6.1 says that $f(x) - \lambda \cdot g(x)$ takes its minimum on W at p. Since $\lambda \cdot g(p) = 0$, this minimum value is $f(p)$.

The last part of the proof is a repeat of the last part of the proof of Theorem 6.3, but we will give the argument for clarity. Let $v \in W$ with $g(v) \leq \mathbb{O}$. Then $\lambda \leq \mathbb{O}$, so that $\lambda \cdot g(v) \geq 0$ and $f(v) - \lambda \cdot g(v) \leq f(v)$. Now compute

$$f(p) = f(p) - \lambda \cdot g(p) \leq f(v) - \lambda \cdot g(v) \leq f(v)$$

This proves that p gives the minimum. ∎

◇ **Problem 129**

Find the local minimums for $y^2 - 2xy + x^3 + x^2 - x$ such that $x^2 + y^2 \leq 2$ and $x \geq -1$. (Note: the case $x^2 + y^2 = 2$ and $x > -1$ is tough; show that the Lagrange Multiplier for the circle has to be positive.)

◇

Bibliography

[1] George Dantzig, *Linear Programming and Extensions*, Princeton: Princeton University Press, 1963.

[2] H. W. Kuhn and A. W. Tucker, Nonlinear Programming, *Proceedings of the Second Berkeley Symposium on Mathematical Statistics and Probability*, pp.481-492, 1950.

[3] Alan Parks, *Introduction to Differential Equations and Linear Algebra*, Appleton: Lawrence University, 2010.

[4] Alan Parks, *Introduction to Real Analysis*, Appleton: Lawrence University, 2014.

[5] Walter Rudin, *Principles of Analysis*, McGraw-Hill, 1976.

[6] Akira Takayama, *Mathematical Economics*, Dryden Press, 1974.

[7] Philip Wolfe, The Simplex Method for Quadratic Programming, *Econometrica*, 27(3) pp.382-98, July 1959.

Index

$A[i, j]$, matrix entry, 1
$A \leq B$ for matrices, 1
A^T, matrix transpose, 5
A^{-1}, matrix inverse, 4
$B(x, r)$, open disk, 70
C^1, continuous partial derivatives, 83
C^2, continuous second partials, 124
$D(x, r)$, closed disk, 76
Df, derivative of f, 81
I_n, identity matrix, 2
P_n, probability space of dimension n, 58
Δ bound, 71
★, spreadsheet problem, 27
∘, dot product of vectors, 18
$\mathbb{D}f$, varied derivative, 84
$\mathbb{H}$, the Hessian, 124
$\mathbb{O}$, a zero matrix, 2
$\mu(A)$, minimum on the sphere, 20
$\nu(A)$, maximum on the sphere, 20
≻, lex-top order for the Simplex Algorithm, 35
$|M|$, norm of matrix M, 18
$|v|$, norm of vector v, 18
$m \times n$, the size of a matrix, 1

activity analysis, 114
affine function, 81, 117
allowable perturbation, 53

basic variables, 11
basic vector, 12
Bolzano-Weierstrass Theorem, 77
boundary, of a set, 76
bounded, set, 76

canonical form, 30
canonical simplex algorithm, 34
Cauchy-Schwarz Inequality, 18
closed disk, 76
closed, subset of n-space, 75
clump, of variables, 89
Cobb-Douglas function, 134
coefficient matrix, linear equation, 5
coefficient matrix, LP, 25
complementary slackness, 51, 105
concave function, 123, 134
constraint complex, 99
continuous, 71
continuous, at a point, 71
continuous, on a set, 71
convex program, 131

www.ingramcontent.com/pod-product-compliance
Lightning Source LLC
Chambersburg PA
CBHW081309170526
45166CB00011B/3465